Multinationals in Latin America

Multinational enterprises, both from industrial countries and from countries within the region operate extensively throughout Latin America. These firms are greatly sensitive to political and economic conditions particularly in less developed countries. The concerns have been well-founded, as waves of nationalism in the 1960s and 1970s led to more and more restrictions on foreign multinationals in Latin America, while the foreign debt crisis of the 1980s has resulted in much greater opening of the regulatory environment to these firms.

This book provides a comprehensive view of the managerial and regulatory issues that relate to multinational firms in the Latin American region. The author discusses issues such as government–business relations as a general concern of multinational firms in the region and also in several specific contexts. The major line of reasoning that runs through the analysis is a bargaining perspective on the multinational enterprise. This view sees the MNE as one of several competing actors for economic and political power in any country; the status of the firm at any time depends on its bargaining with the other actors for economic resources and a political base. Clearly, both of these ideas play important parts in the activities of MNEs in Latin American countries.

This perspective is followed through a range of managerial subjects — overall business strategy, organizational structure, and financial management — showing how common business decisions are affected by the bargaining relationship in Latin America.

The author:
Robert Grosse is Director of the International Business and Banking Institute at the University of Miami.

MULTINATIONALS IN LATIN AMERICA

Robert Grosse

London and New York

First published 1989
by Routledge
11 New Fetter Lane, London EC4P 4EE

Simultaneously published in the USA and Canada
by Routledge
a division of Routledge, Chapman and Hall, Inc.
29 West 35th Street, New York, NY 10001

Reprinted 1990

Phototypeset in 10pt Times by
Mews Photosetting, Beckenham, Kent

© 1989 Robert Grosse

Printed in Great Britain by
Antony Rowe Ltd, Chippenham, Wiltshire

British Library Cataloguing in Publication Data

Grosse, Robert E.
   Multinationals in Latin America —
   (International business)
   1. Latin America. Multinational companies
   I. Title II. Series
   338.8′888
   ISBN 0-415-00398-9

Library of Congress Cataloging-in-Publication Data

Grosse, Robert E.
   Multinationals in Latin America / Robert Grosse.
     p. cm. — (International business series)
   Includes index.
   ISBN 0-415-00398-9
   1. International business enterprises — Latin America.
   2. Investments. Foreign — Latin America. I. Title. II. Series:
   International business series (London, England)
   HD2810.5.G76   1989
   388.8′888-dc19   88-30328
                                           CIP

To my parents,
for their never-failing support

Contents

Contents

Figures

Tables

Acknowledgements

I would like to thank my research assistant, Mr Diego Aramburu, for his careful and thoughtful work in carrying out statistical tests and tracking down information related to this project. In addition, I would like to thank the many company managers who contributed their time and efforts to provide information for the various surveys used throughout the book. In particular, the following executives gave freely of their time to help with preparation of the book: John Carnegie, Yves Bobillier, Harry Schiefiele, Tony Tremols, Bernie Hamilton, and Thomas Layman. Several academic colleagues also offered their comments and assistance in parts of the project. They include: Alan Rugman, Greg Smogard, Marianela Hernandez, and Jack Behrman.

Introduction

This book is about multinational enterprises (MNEs) that operate in Latin America. It deals primarily with MNEs from industrial countries such as the United States and the United Kingdom, though locally-based private firms and state-owned MNEs also are considered. The level of analysis is microeconomic, that is, it focuses on decision-making within these firms, on public policy used to direct the activities of these firms, and on competition among them. This book is not a condemnation of multi-national enterprises for their inability to solve Latin America's economic development or income distribution problems, nor is it a celebration of the virtues of such firms. Essentially, the book represents an attempt to explain what the (primarily foreign) MNEs in the region do and why they act as they do.

The basic theme of this volume is a view of multinational business that considers the MNEs as a goal-oriented actor (seeking profits and growth), constrained by acts of government policy-makers and competing firms, each seeking somewhat conflictive goals. This conceptual framework is based on a bargaining theory of the MNE that is explained in Chapter 2 and substantially linked to subsequent chapters.

The conceptual framework is as follows. The MNE functions as to increase its own value, dealing with regulators and competitors to obtain the best position for itself under the prevailing conditions in each country. More precisely, the MNE follows the strategy of *internalization*, to take advantage of its competitive strengths, and it operates under the 'bargaining theory of the MNE,' trying to gain the most beneficial results in dealing with environmental factors (especially competitors and regulators). The idea of internalization has been explored by Rugman (1981) and others, and needs no clarification here. The bargaining theory of the MNE is laid out in Chapter 2 and tested and extended later. The perspective is reasonably similar to that of Gladwin and Walter (1980). In simple terms, it views the MNE as one of several competing actors for economic and political power in any country; the status of the firm at any time depends on its bargaining with the other actors for economic

1

resources and a political base. Clearly, both of these ideas play important parts in the activities of MNEs in Latin American countries.

Part I lays out the historical and theoretical background that forms the underpinning of the book. Following a historical survey of multinational business in Latin America (Ch. 1), the conceptual framework is laid out (in Ch. 2). Chapter 3 then discusses current conditions and trends facing MNEs in the region.

Part II focuses on the regulatory setting and the macroeconomic environment that broadly characterize the region today. Clearly, with the debt crisis of the 1980s, Latin American governments have become more open to foreign MNE participation in their countries, although the depressed economies of all countries have discouraged multinational business there. Chapter 4 looks at the government–business relationship. With elected governments in almost all countries of the region (with major exceptions in Chile and Nicaragua), and greatly diminished confidence in the import-substitution strategy of development, the regulatory environment has improved notably during the current decade. One key issue that remains actively debated and subject to important change in the near future is the view of the proper roles of public and private sectors held by Latin American governments — a view which differs essentially country by country in the region. A second, related issue is the use of government-owned companies in each country, since these companies typically compete with MNEs (or preclude them from operating in some sectors). Some of the main laws on multinational business in selected countries are discussed as well. Finally, related to the bargaining theory, the issue of the 'obsolescing bargain' is discussed. After presenting these and other perspectives on the government–business relationship, the bulk of the chapter is used to test the bargaining theory of the MNE.

Chapter 5 follows the broad discussion of government–business relations with an attempt to illuminate the key issues of concern to governments when they evaluate foreign direct investment projects. This chapter discusses the major categories of economic impact that are considered by host governments, and it presents findings of a study of foreign manufacturing MNEs in Venezuela during 1984–5.

Chapter 6 uses the bargaining theory to explore the issue of transnational regulation of MNEs, as carried out specifically in the Andean Pact countries. This group of countries used a particularly restrictive set of rules on foreign direct investment from 1971 to 1987 that demonstrates both the power positions of companies and governments, and also the shifting regulatory tide that has swept through the region since 1971.

Completing this part, Chapter 7 analyzes the Latin American debt crisis, which has been the single most important phenomenon affecting business of all kinds in the region during the 1980s. The chapter explores the roots of the crisis, the particular situations of Argentina, Brazil,

Mexico, and Venezuela today, and prospects for the next five years. Part III turns to managerial issues — strategy, finance, and organization. Each chapter of this part focuses on one functional area and on one empirical study. Chapter 8 is a study of those Fortune 500 companies that operate relatively extensively in Latin America. Using a questionnaire survey, the analysis attempts to correlate company characteristics with measures of performance in the region. The conceptual base of the analysis is a theory of competitive advantages that explains how firms will operate in different competitive environments. This view differs from the main line of reasoning in the book by focusing on inter-company rivalry rather than on company–government relations.

Chapter 9 presents results of an empirical study performed in 1981 in Peru. This study looked at foreign MNEs' strategies (and organizational structures) for their operations in this country. The inter-industry differences in strategy and structure are highlighted. For example, oil firms have shifted to long-term contracting rather than ownership of oil–producing subsidiaries; manufacturers have continued to use owned subsidiaries for their business, though often with local joint-ventures partners; and banks have followed very different strategies in different countries, due to extensive and varying regulation.

Chapter 10 examines the finance function, with particular emphasis on methods available to MNEs for moving funds (or other instruments of financial value) from one country to another (e.g. from a Latin American country back to the home office, or the reverse). The typical, restrictive financial rules in Latin American countries preclude or greatly hinder traditional means of inter-company transfers such as dividends, loans, and royalties, making other vehicles more important in this context. The empirical evidence comes from a survey of home offices and regional offices of US–based MNEs that operate subsidiaries in Latin America.

Part IV illustrates the book's broad view of MNE business in Latin America with a look at two specific kinds of MNEs in the region– chemical companies and government-owned MNEs — and concludes with a look to the future.

Chapter 11 investigates the chemical industry. A small group of major mutinational chemical companies with extensive operations in Latin America form the base for this analysis of competitive conditions and government relations in the industry. The main points of this chapter are to illustrate the broad issues raised throughout the book in the context of one set of firms and to demonstrate how the 'industry' is actually a set of competitive groups that compete differently in each country and segment.

Chapter 12 deviates from the previous sections by concentrating on

Introduction

Latin American MNEs, specifically the government-owned firms in industries such as petroleum, transportation, and banking. These firms tend to be very large in their home countries, but generally relatively small in comparison with the large MNEs from North America, Europe, and Japan. These firms are a central concern for foreign MNE managers, since they are often large, and financially strong, competitors in the host-country market and abroad. The success or failure of this business form in the coming years will have a tremendous impact on the role allowed to foreign MNEs in less-developed countries, especially in Latin America. Examples of such firms in the region include major oil companies (PDVSA, Pemex, Petrobras), mining companies (Mineroperu, Codelco), and banks (Banco de la Nacion, Banamex, Banco do Estado de Sao Paulo). Chapter 12 gives an overview of the activities, performance, and outlook for state-owned MNEs from Latin American countries.

Chapter 13 concludes the book, with a review of the contents and a look to the future.

Bibliography

Gladwin, Thomas, and Ingo Walter (1980) *Multinationals under Fire*, New York: Wiley.
Rugman, Alan (1981) *Inside the Multinationals*, Beckenham: Croom Helm.

4

Part one
Historical and Theoretical Background

Chapter one

A history of MNE activities in Latin America

Introduction

Sources of external finance in Latin America

The major form of MNE business, that is foreign direct investment (FDI), began long before the independence of Latin American countries from European colonial powers. Spanish and Portuguese individuals and firms owned a variety of investments — such as farms and raw materials ventures — in the Latin American colonies before 1800.[1] In order to make the task of this chapter manageable, the story will be taken up just after the region's independence wars, that is in the early 1800s. This introductory section offers an overview of long-term investment in Latin America during the period up to the Second World War. Subsequent sections look in more detail at the activities of British and American firms in the region from 1820 to 1980. The final section discusses MNEs from other countries operating in Latin America since the Second World War, and then offers some conclusions concerning trends in MNE activities and in government–business relations. Discussion of the 1980s is left for Chapter 3 and subsequent chapters that focus on specific MNE-related business issues in the region.

The colonial period in Latin America largely came to an end in the first two decades of the 1800s, when Argentina, Brazil, Chile, Colombia, Peru, and several other countries won their independence from Spain, or in Brazil's case, from Portugal. The earliest major form of long-term capital flow into Latin American countries was the purchase of bonds issued by Latin American governments soon after their independence battles. Most of these purchases (of about £20 million in value) were made by British merchant banks and other financial intermediaries in the London market. In addition, some direct investment (of about £4 million in value) took place in companies established to explore for gold and silver in former colonies such as Mexico, Peru, and Chile.[2]

The bonds were issued in the London financial market by investment banks on behalf of Latin American governments, which in turn used the funds largely to pay for war debts incurred previously. The bonds were quite speculative, selling at substantial discounts under face value and paying high rates of interest. When the Latin American governments became unable to service their interest payments by the late 1820s because of inadequate export earnings, the market for the bonds collapsed and remained inactive for over two decades. (The parallel with the 1980s is quite striking, if for 'government bonds' we read 'sovereign debt' to commercial banks!) Similarly, the investments in mining company shares were largely lost to bankruptcies, though a few of the mines continued to produce ores profitably for many years.[3] It was not until the Industrial Revolution began to spread to Latin America that a new wave of investment entered the region. Table 1.1 sketches the early British long-term investments, both loans and stock/bond purchases. Government loans remained stagnant until the 1860s, with primarily refinancing of defaulted loans as the main area of increased lending. Investment in mining ventures actualy grew during the 1825–40 period, despite the bankruptcies, and then declined until 1865. In fact, the real growth in investment came from the arrival of the Industrial Revolution.

Table 1.1 British investment in Latin America, selected years (£000s)

Investment	1825	1840	1865
Government loans*	20,749	23,580	61,781
Mines	3,570	6,475	6,208
All other	245	752	16,381
Total	24,564	30,807	80,870

Source: Stone 1968: 315.
*Outstanding principal.

By the 1850s the advent of railroads and telegraph led to major investments in Latin America, again primarily from the United Kingdom. Both loans to national governments (in the form of bond issues on the London Stock Exchange) and direct investment in the railroad and utility companies increased British investment in the region substantially. Britain's leadership in the Industrial Revolution gave British firms the competitive advantages that could be exploited through direct investment, contracting with local firms, and otherwise transferring the industrial knowledge gained in the United Kingdom to Latin America. While the early direct investments primarily served the local market needs for transportation and communication, they also later served to facilitate dealings with foreign purchasers of raw materials and suppliers of imports — as well as enabling MNEs to communicate with local affiliates more easily and efficiently.

It should be recognized at this point that most of what is being termed 'direct investment' in the nineteenth century was quite different from the common perception of this activity in the period since the Second World War. In the nineteenth century much of the direct investment that entered Latin America was indeed undertaken by British joint-stock companies, but these firms were generally formed by risk-taking investors to support the entrepreneurial efforts of a few British expatriates who emigrated to the New World. These expatriates, in turn, hired local workers and built local companies to operate railroads, telegraph companies, and mines in Latin America. Thus there was not a transfer of knowledge and skills that is typical of the modern MNE, but rather a transfer of funds that enabled British entrepreneurs to start essentially new firms and operate them. In terms of ownership, the investment was in fact direct. In terms of control, while it is true that the Europeans controlled the purse strings, it is more accurate to say that operating control tended to rest with the expatriates in Latin America. This is significantly different from the 'multinational enterprise' type of foreign direct investment that occurs today, when domestic and foreign activities are (at least somewhat) integrated parts of a total firm. [This point was suggested by Mira Wilkins, who graciously reviewed the present chapter.]

Overall, the picture of long-term capital flows (direct and portfolio investments) to Latin America during most of the 1800s showed primarily British participation, and was dominated by lending to governments through bond issues and direct investment rather than commercial bank lending or government lending. The United States was a minor factor in the capital flows until the end of the century, and indeed the United States still was a net borrower in international financial markets during that period. Table 1.2 depicts some of the broad characteristics of investment in the nineteenth century.

By the turn of the century, the nature of foreign investment in Latin America was moving more toward the private sector and away from government bonds. Bond sales in the London market still remained substantial, but the issuers were railways, mining companies, and other private-sector firms. Direct investment had become predominant in relation to the portfolio investment in bonds for British investors. Investors from the United States also were placing much more of their funds into direct investments. In 1908 the stock of US investment in the region included $334 million of portfolio investment and $749 million of direct investment (primarily in railroads and mining).[4] (Foreign aid, another source of foreign financing that typically figures in twentieth-century measures of foreign capital entering Latin America, was virtually non-existent during the period up to the Second World War.)

As the twentieth century began to unfold, the United States gained in industrial power relative to the United Kingdom. In Latin America

9

Table 1.2 Long-term capital flows into Latin America in the 1800s

Source & date	Long-term bond issues	Foreign direct investment
1822–5 United Kingdom	£21 million by London banks for various Latin American governments	More than 40 joint-stock companies formed to extract raw materials in the region
1825 France	FF30 million issued by the govt of Haiti on the Paris Stock Exchange (Bourse)	
1826–50 United Kingdom	£18 million by London banks for various Latin American governments	No new joint-stock firms founded to operate in Latin America
1851–80 United Kingdom	£130 million by London banks for various Latin American governments	Railways built in several countries, including Peru and Brazil
1870–9 France (annual average)	FF1.5 billion issued by various Latin American govts on Paris Bourse	FF32 million of private-sector shares and bonds issued on Paris Bourse
1881–1900 United Kingdom	£105 million net increase in Latin American bond issues on London Stock Exchange	£147 million FDI in railways during 1880–90; other private stock and bond investment rose by £256 million during 1880–1900
1880–99 France	FF2.1 billion issued by various Latin American govts on Paris Bourse	FF2.3 billion of private-sector shares and bonds issued on Paris Bourse

Source United Nations 1965: 5–12.

this meant that US lenders (bond purchasers) began to replace British ones, and US direct investors began to seek out raw materials' supplies in the region. By the end of the 1920s, US capital flows exceeded British flows to Latin America, and the total stock of investment had become predominantly American. Direct investment outweighed portfolio investment from both sources. Estimates of the external financing from both countries in the early twentieth century appear in Table 1.3.

Table 1.3 Foreign investment in Latin America during 1900–39

Year & country	Public-sector foreign indebtedness	Private-sector foreign indebtedness (stocks & bonds)
1900		
United Kingdom	£228 million	£312 million
France	FF2.46 billion	FF505 million
United States (1897)	n.a.	$304 million
1913		
United Kingdom	£317 million	£683 million
France	FF5.12 billion	FF3.04 billion
United States (1914)	$366 million	$1.28 billion
1928		
United Kingdom	£341 million	£858 million
France	n.a.	n.a.
United States (1929)	$1.72 billion	$3.46 billion of FDI
1939		
United Kingdom	£179 million	£579 million
France	n.a.	n.a.
United States (1940)	$1.57 billion	$2.70 billion of FDI

Countries of origin of capital flows to Latin America

The United Kingdom and the United States were the sources of at least two-thirds of long-term capital flows to Latin America during the entire period up to the Second World War. As already noted, the United Kingdom was the first major creditor country to Latin American nations that won independence from Spain or Portugal in the early 1800s. During the entire century, UK investors played the largest role in external financing of Latin American economies and in the transfer of industrial knowledge through foreign direct investment.

France, Germany, and the United States subsequently increased in importance as suppliers of foreign capital (and knowledge) to the region. By 1914, US firms, individuals, and governments had become major investors in the region, surpassing all source countries except the United Kingdom. Table 1.4 gives some details concerning the distribution of investors and recipients of foreign *direct* investment stocks in 1914. Note the heavy concentration of investments in the three largest countries (Argentina, Brazil, and Mexico) and the dominance of British and American investors overall. French direct investment was concentrated in Argentina and Brazil at that time, and it far exceeded US investments in those countries. In contrast, US investment was much heavier in neighboring Mexico, and in Central America. UK investments were located much more in South America and particularly in Argentina. Not only did British investment dominate the total (at close to 50 per cent),

11

but this was a major part of total British overseas investment at the time. The $US3.6 billion total British long-term investment in Latin America was approximately the same magnitude as total British long-term investment in the United States.[5]

Table 1.4 Latin America: foreign private investment at the end of 1914 (US$m)

Debtor countries	Creditor countries					
	United Kingdom	France	Germany	United States	Others	Total
Argentina[b]	1,502	289	235	40	1,151	3,217
Bolivia[c]	17	25	. . .	2	. . .	44
Brazil	609	391	. . .	50	146	1,196
Chile[d]	213	. . .	56	225	. . .	494
Colombia	31	1	. . .	21	1	54
Costa Rica	3	. . .	. . .	41	. . .	44
Dominican Republic	0	. . .	. . .	11	. . .	11
Ecuador	29	2	. . .	9	. . .	40
El Salvador	6	. . .	. . .	7	2	15
Guatemala	44	. . .	12	36	. . .	92
Haiti	0	. . .	. . .	10	. . .	10
Honduras	1	. . .	. . .	15	. . .	16
Mexico	635	. . .	. . .	542	. . .	1,177
Nicaragua	2	. . .	. . .	4	. . .	6
Panama	0	. . .	. . .	23	. . .	23
Paraguay	18	. . .	. . .	5	. . .	23
Peru[d]	121	1	. . .	58	. . .	180
Uruguay	154	. . .	2	0	199	355
Subtotal	*3,385*	*709*	*305*	*1,099*	*1,499*	*6,997*
Venezuela	30	2	15	38	60	145
Subtotal II	*3,415*	*711*	*320*	*1,137*	*1,559*	*7,142*
Cuba	170	. . .	. . .	216	. . .	386
Subtotal III	*3,585*	*711*	*320*	*1,353*	*1,559*	*7,528*
Undistributed by debtor country	—	—	—	41	—	41
Total	*3,585*	*711*	*320*	*1,394*	*1,559*	*7,569*

Source: United Nations 1965, Table 17.
Notes:
[a] The figures in European or Latin American currencies have been converted into dollars on the basis of the gold exchange par at the end of 1914. The figures correspond to the amounts outstanding on 31 December unless otherwise indicated.
[b] Outstanding amount as of 31 December 1918.
[c] Outstanding amount as of 31 December 1917.
[d] Outstanding amount as of 31 December 1915.
. . . Values not available.

After the First World War, European investment in Latin America diminished greatly as the Allies rebuilt their own economies from war damages, and Germany had many investments confiscated by national authorities in the region. US investment increased somewhat, but it appears that total foreign investment in Latin America decreased from 1914 to 1929. The US share increased from about 17 per cent of total foreign long-term capital in 1914 to about 40 per cent in 1929.[6] During the Depression of the 1930s, very little new foreign investment entered the region. External debt throughout the region was defaulted beginning in 1931, and so new bond issues in New York or London were not viable. This debt crisis was due partially to what in retrospect appears to be overborrowing during the late 1920s and partially to steep declines in the prices of Latin America's main export commodities: sugar, copper, precious metals, and petroleum. The stock and flows of foreign direct investment declined as well, due to diminishing market capacities of the countries and also to increased restrictions on the transfer of earnings abroad.

The Second World War brought with it an increased demand for Latin America's commodities, especially by the United Kingdom and the United States. This stimulus to exports, and a much slower increase in importing, led to favorable balance-of-payments positions for most of the countries of the region during the war. Given the industrial countries' war efforts, however, the improved economies in Latin America still did not attract noticeable increases in foreign investment until after 1945.

Target industries of foreign investment in Latin America

Throughout the first 150 years of post-colonial experience in the region, foreign investment was concentrated in natural-resource industries and public utilities (e.g. power generation, telephone and telegraph service, and public transportation). Construction and operation of railroads in the mid-1800s attracted both foreign debt and equity investments which far surpassed any other industry. By the end of the century, in both US and UK direct investments, railroads accounted for about half of the total. Additional targets of British investment included public utilities and banking, but the combined shares of these industries in total British investment never exceeded 10 per cent. Direct investment by the US firms showed a rapid decline in railroads after 1900, and increased shares of mining and agriculture. Table 1.5 shows the sectoral distribution of US direct investment form 1897 to 1929. Note that by the time of the Depression, US direct investment had shifted to sugar, oil, mining, and public utilities (especially telephone companies, that ITT and other US firms took over when British firms divested). Railroad investment was

Table 1.5 United States: direct investment in Latin America, by sector, 1897–1929 (aggregate totals in US$m at end of year)

Branch of economy	1897		1908		1914		1919		1924		1929	
	Total	%	Total	%	Total	%	Total	%	Total	%	Total	%
Agriculture	56.5	18.6	158.2	21.1	238.5	18.7	500.1	25.3	830.6	30.0	877.3	24.1
Sugar	24.0		57.0		115.0		354.0		668.0		643.5	
Fruit	8.5		28.2		57.5		67.1		86.1		153.8	
Others	24.0		73.0		66.0		79.0		76.5		80.0	
Mining and smelting	79.0	26.0	302.6	40.4	552.2	43.3	660.8	33.4	713.0	25.7	801.4	22.0
Precious ores and stones	58.0		141.6		176.2		145.5		151.0		164.3	
Industrial minerals	21.0		161.0		376.0		516.3		562.0		637.6	
Petroleum	10.5	3.5	68.0	9.1	130.0	10.2	326.0	16.5	533.0	19.2	731.5	20.1
Production	3.5		55.0		107.0		286.0		473.0		654.0	
Distribution	7.0		13.0		23.0		40.0		60.0		77.5	
Railroads	129.7	42.6	110.0	14.7	175.7	13.8	211.2	10.7	261.3	9.4	230.1	6.3
Public utilities	10.1	3.3	51.5	6.9	98.4	7.7	101.0	5.1	161.9	5.8	575.9	15.8
Manufacturing	3.0	1.0	30.0	4.0	37.0	2.9	84.0	4.2	127.0	4.6	231.0	6.3
Trade	13.5	4.4	23.5	3.1	33.5	2.6	71.0	3.6	93.0	3.3	119.2	3.3
Others	2.0	0.6	5.0	0.7	10.5	0.8	23.5	1.2	59.5	2.0	79.4	2.2
Total	304.3	100.0	748.8	100.0	1,275.8	100.0	1,977.6	100.0	2,779.3	100.0	3,645.8	100.0

Source: United Nations 1965, Table 15.

minimal by 1929, and indeed it continued to decline subsequently. In all of these industries except railroads, the foreign firms were able to compete on the basis of their ability to obtain large amounts of capital and their access to large foreign markets to sell the raw materials. During the Depression, total US direct investment in Latin America fell slightly (from a total of about $3.5 billion in 1929 to about $2.7 billion in 1940), and its sectoral distribution changed significantly. Declines in agriculture and mining were offset by increases in petroleum and manufacturing. In the railroad industry (as well as in several others) US direct investors replaced many European investors during the Depression; so the total amount of direct investment in railroads remained about constant from 1929 to 1945.[7]

The next two sections focus respectively on UK and US direct investments and MNE activities in Latin America over the past century and a half.

The UK experience, 1820–1980

Before the Second World War

During the first half of the nineteenth century, British direct investment was fairly small relative to portfolio investment. The forty-six joint stock companies incorporated on the London Stock Exchange during 1822–5 to do business in Latin America mostly were formed to undertake mining ventures. The General South American Mining Association was the largest company, with capital of £25 million.[8] As previously mentioned, because of defaults on most of the government bonds issued in the 1820s on the London Stock Exchange for Latin American government borrowers, new long-term investment in the region dried up until about 1850. In addition, most of the joint-stock companies went bankrupt during the 1820s, and no new ones were set up for about twenty-five years.

As one of the first steps in transferring the Industrial Revolution to Latin America, transportation and other infrastructure were developed by British firms and individuals. Railroads were constructed in several Latin American countries by British investors beginning in 1849. In that year railroad companies were incorporated to construct rail lines and operate trains in Panama and Brazil. Additional companies followed in Argentina, Chile, Mexico, and Venezuela during the following decades.[9] While the railroad companies were able to establish themselves and begin operations on the basis of foreign technology and capital brought into a country, continued survival depended on successful ongoing dealings with the host government. The companies apparently did succeed in such dealings well into the twentieth century, judging from

15

the record of profitability of the railroads and the small number of nationalizations until after 1900.

By 1890 there were sixty-nine British-owned mining companies and twenty nitrate firms operating in the region, along with the railroads, agricultural projects, and various other ventures.[10] Manufacturing ventures numbered only thirteen in the entire region at that time, including tobacco factories in Cuba, sugar refineries and a meat-processing plant in Brazil, and four meat-processing plants in Argentina. Despite the diverse sectoral distribution of long-term British investment, virtually all of it was concentrated in shares and bonds of railroad companies (39 per cent) and bonds of national governments (46 per cent).[11]

During the first fifteen years of the twentieth century, British investment in the region doubled, with most of the new funds still going into railroads, public utilities, and mining ventures. Beyond these sectors, new investment began to build up in petroleum extraction and refining; Royal Dutch-Shell Company was exploring, producing, or selling in almost all Latin American countries by that time. Also the banking industry attracted substantial British investment, especially by major commercial banks such as Barclays and the London Bank of South America. At the beginning of the First World War, British investments accounted for approximately two-thirds of total foreign investment in Latin America.[12]

The sectoral distribution of British investment in Latin America during the period from the mid-1800s to the First World War is summarized in Table 1.6.

In terms of geographical distribution, British investment extended from Mexico to Argentina, although during the last quarter century before the First World War it became more and more concentrated in the largest countries, namely Argentina, Brazil, and Mexico. By 1913 there were about 620 British-owned direct investments in Latin America, including 118 in railroad ventures (which constitutes 64 percent of total direct investment in value).[13]

After the First World War, British investment did not increase noticeably. In fact, as has been mentioned, by the time of the Depression the United States had replaced the United Kingdom as Latin America's largest source of foreign investment. Over the course of the Depression, British direct investment declined steadily, to about £800 million in total. British portfolio investment shifted from large holdings of government bonds in the late 1800s to only 28 percent government debt by 1928. Given the widespread government bond defaults during the Depression, and the subsequent redemption of many of those issues (at often 10–15 percent of face value) in the late 1930s, government debt became a very small part of total British investment in Latin America by 1940.

Table 1.6 Industrial composition of British investment in Latin America, 1865–1913 (£000s)

	1865	1875	1885	1895	1905	1913
Government loans	£61,781	£129,360	£161,160	£262,377	£307,760	£445,481
Railroads	9,551	24,085	55,184	199,926	237,288	404,535
Public utilities						
Canals & docks	—	116	201	457	1,760	15,036
Electric light & power cos.	—	—	—	—	1,207	24,915
Gas companies	848	1,919	2,443	2,250	5,003	7,166
Telegraph & telephone cos.	—	5,380	5,614	7,699	7,199	9,953
Tramways & omnibus cos.	—	991	1,539	5,383	23,243	79,185
Waterworks	—	—	703	2,012	1,823	2,837
Financial						
Banks & discount companies	2,012	3,201	2,350	5,023	9,424	24,252
Financial, land & invest. cos.	—	151	1,804	34,479	30,461	57,104
Financial trusts	—	—	—	—	10,979	13,090
Raw materials						
Coffee & rubber companies	—	—	—	—	1,929	4,703
Mines	2,708	2,456	7,831	11,676	15,644	16,613
Nitrate companies	—	48	—	6,697	9,869	12,392
Oil companies	—	—	—	—	171	4,477
Industrial & miscellaneous						
Breweries & distilleries	—	—	—	1,218	1,046	1,165
Commercial & industrial	1,044	1,992	4,707	10,158	17,481	34,569
Iron, coal & steel	—	90	130	56	—	1,662
Shipping companies	2,926	4,820	2,953	3,095	5,981	18,330
Total	£80,869	£174,611	£246,620	£552,505	£688,268	£1,177,462

Source: Stone 1868: 323.
Note: Details may not add due to rounding.

The sharpest decline in British direct investment in the region occurred during the 1940s. First, the war disrupted British business overseas and led to divestments in Latin America. Second, immediately after the war, Latin American governments purchased railways and other infrastructure projects from British investors. These factors left British direct investment in the region at a low level of £238 million by 1949.[14]

After the Second World War

During the period from 1950 to 1980, direct investment by British firms grew consistently in the region as a whole, with geographic concentration remaining in the Southern Cone countries and Mexico. Oil and utility investments that had been purchased by local governments were replaced by manufacturing and service investments. While British investment was growing, German, Swiss, and Japanese direct investment grew much faster, such that by 1980 the United Kingdom had been surpassed in total Latin American direct investment by Germany and was being challenged by Japan as well. Despite this turn of events, the presence of British firms is still very evident in the region. And the Royal Dutch-Shell network of oil exploration, refining, and marketing affiliates remains by far the largest non-US direct investment in the region.

The US experience, 1820–1980

Before the First World War

The period up to 1900 was characterized by the United Nations study as follows:

> [US] Direct investment included only some silver and gold mines in Mexico [e.g. ASARCO and the Guggenheim brothers], a few sugar plantations bought by American trading companies in Cuba during the political disturbances of 1876–78 [e.g. Drake Brothers and Company, American Sugar Refining Company], a railroad in Panama [the Panama Railroad Company, owned by the Pacific Mail Steamship Company] and some trading firms in Peru [e.g. W.R. Grace and Company], Argentina [e.g. Central and South American Cable Company], and Colombia [e.g. United Fruit Company].
>
> (United Nations 1965: 13)

While Wilkins (1970) disagrees, documenting dozens of small US direct investments in Latin America, still the value of this investment outside of the ventures noted above was very small. She notes the extensive, though small-scale operations of the Singer Sewing Machine Company, several telephone and telegraph companies, Standard Oil of New Jersey, the Boston Fruit Company (which became United Fruit Company in 1899) and Edison Electric Company in the region (ibid., Ch. 3). One type of investment missed by the UN study that should not be overlooked were the dozens of railroads set up across Mexico by US firms. By 1897 there was over $100 million of railroad investment by American companies in Mexico — the single largest sector for US direct investment in the single largest host country at that time.[15]

From only $635 million in 1897, US direct investments in Latin America grew to $2.7 billion in 1914.[16] According to a study by Cleona Lewis (1938), most of the investment went into agricultural ventures in Cuba and into railroads, mining, and petroleum in Mexico. Some 77 percent of all US investment in the region was in Central America, the Caribbean, and Mexico. In fact, Mexico and Cuba accounted for 85 percent of this investment (and 66 percent of total US investment in Latin America). The sectoral disaggregation of US direct investment at the time was shown above in Table 1.4.

The concentration of direct investments in countries nearby to the United States and the emergence of the United States as the leading industrial power in the world contributed to US government policies that have subsequently become known as 'gunboat diplomacy.' In order to protect US firms doing business in Cuba, Nicaragua, and Haiti, US marines were sent to guard US citizens and properties in 1906, 1910, and 1914 respectively. In fact, 'from 1912 to 1933 (except for a brief trial interlude in 1925) a small force of marines remained in Nicaragua to keep stability.'[17] This policy presumably ended after the Nicaraguan episode, but the threat of US intervention has created political problems for US firms in Latin America ever since.

1914–1939

During the First World War, US firms sought out additional supplies of raw materials in Latin America for the war effort. The major meat-packing companies — Swift, Armour, Morris, and Wilson — all invested in plants in Argentina and other Southern Cone countries. Similarly, sugar companies, such as Amalgamated Sugar, expanded their producing, milling, and transporting operations in Cuba. Mining ventures proliferated in the Andean countries, where tin (Bolivia), copper (Peru and Chile), iron ore (Chile), and bauxite (Guyana and Dutch Guinea) were extracted. Of course, petroleum exploration and extraction were intensified in the region — especially Mexico, Peru, and Central America — (by Standard Oil of New Jersey and Sinclair Oil, among others).[18] And following the industrial firms, two of the largest commercial banks began to set up branch networks throughout Latin America; First National City Bank opened in 1914 in Buenos Aires, and by 1917 had offices in Brazil, Uruguay, Chile, and Cuba. First National Bank of Boston similarly opened first in Buenos Aires (in 1917) and later expanded throughout the region.

After the war ended, US direct investors continued to expand their operations in Latin America, often buying existing companies from European firms that suffered from the war and needed to retrench their operations back in Europe. For example, the Guggenheim family (which held major stakes in Kennecott and ASARCO copper companies) acquired the British Anglo-Chilean Nitrate and Railway Company in 1924. Another major example was the American & Foreign Power Company, originally part of General Electric Company. This firm initially obtained from GE plants that provided power and lighting and operated street cars in Panama, Guatemala, and Cuba. During the 1920s, it purchased

Table 1.7 Latin America: US direct investment by country, 1929–50 (US$m)

Country	1929	1936	1940	1943	1950
Caribbean countries	*1,002*	*717*	*612*	*611*	*761*
Cuba	919	666	559	526	642
Dominican Republic	69	41	41	71	106
Haiti	14	10	12	14	13
Mexico and Central America	*917*	*628*	*542*	*569*	*727*
Mexico	682	480	357	286	415
Costa Rica	22	13	24	30	60
El Salvador	30	17	11	15	17
Guatemala	70	50	68	87	106
Honduras	72	36	38	37	62
Nicaragua	12	5	8	4	9
Panama	29	27	36	110	58
South America	*1,543*	*1,458*	*1,542*	*1,541*	*2,957*
Argentina	332	348	388	380	356
Bolivia	62	18	26	13	11
Brazil	194	194	240	233	644
Chile	423	484	413	328	540
Colombia	124	108	111	117	193
Ecuador	12	5	5	11	14
Paraguay	11	5	5	9	6
Peru	124	96	81	71	145
Uruguay	28	14	11	6	55
Venezuela	233	186	262	373	993
Total	*3,462*	*2,803*	*2,696*	*2,721*	*4,445*

Source: United Nations 1965, Table 29.

properties throughout Latin America from British, German, French, Canadian, and local investors; and it became the largest public utility company in Latin America (larger than ITT, which also had extensive holdings in several countries).[19] Once the Depression began, US investment of all kinds in Latin America dropped off significantly, and remained low until the Second World War. During 1930, expansion that had begun in the 1920s actually continued to a large degree (e.g. ITT and American & Foreign Power Company both purchased additional Latin American phone companies and other properties, and Pan Am expanded its operations in the region), but by 1931 the trend was to close operations and forego new investment. Across the range of industries involved in Latin American direct investment, the decade of the 1930s brought the same Depression in Latin America that had struck in the United States and other industrial countries.

The overall evolution of aggregate US direct investment in Latin American countries during the Depression (and to 1950) is shown in Table 1.7. With very few exceptions, the stock of direct investment fell in nominal as well as real terms during the period from 1929 to 1940.

1940–1973

From the beginning of the Second World War into the 1960s, direct investment in Latin America (and consequent MNE activity there) followed much the same pattern as during and after the previous war. That is, European and other direct investments dropped off significantly, while US investment grew steadily. Given the relative lack of data on non-US firms, aggregate analysis must rely on US Department of Commerce data concerning US firms only. This is not too overwhelming of a problem, since US investment accounted for well over half of the total since 1940. Table 1.8 shows direct investment flows for selected years and into selected countries. Note that up until the Castro revolution, Cuba was a leading recipient of US direct investment, surpassed only by Venezuela and Brazil. Also note that the oil industry in Venezuela raised that country's attractiveness to FDI far beyond its relative size compared to Brazil, Mexico, or Argentina. In fact, these two phenomena (Cuban nationalizations and Venezuelan oil production) were the causes of the most notable variations in FDI flows into Latin America from the Second World War to 1975. The trend in manufacturing has been fairly steady and increasing during the entire post-war period up to 1982. The next few paragraphs point out some of the significant factors related to MNE activities during each decade up to the 1980s.

Factors that affected the flows of direct investment into the region during the 1940s included the retrenchment of European investors, whose domestic businesses in Europe had been damaged in the war. US-owned

Table 1.8 US investments in Latin America and the Caribbean, 1943–81

Countries	1943 $m	%	1950 $m	%	1964 $m	%	1972 $m	%	1981 $m	%
Total = 1 + 2 + 3 + 4	2,721.0	100.0	4,735.0	100.0	10,205.0	100.0	16,796.0	100.0	38,883.0	100.0
1. *South America*	1,541.0	56.6	2,957.0	62.4	6,612.0	64.8	9,545.0	56.8	18,109.0	46.6
Argentina	380.0	14.0	356.0	7.5	882.0	8.6	1,403.0	8.3	2,735.0	7.0
Bolivia	13.0	0.5	11.0	0.2	997.0	9.8	a	a	a	a
Brazil	233.0	8.6	644.0	13.6	a	a	2,205.0	14.9	8,253.0	21.2
Chile	328.0	12.0	540.0	11.4	789.0	7.7	620.0	3.7	834.0	2.1
Colombia	117.0	4.3	193.0	4.0	508.0	5.0	737.0	4.4	1,178.0	3.0
Ecuador	11.0	0.4	14.0	0.3	a	a	a	a	277.0	0.7
Paraguay	9.0	0.3	6.0	0.1	a	a	a	a	a	a
Peru	71.0	2.6	145.0	3.1	464.0	4.5	712.0	4.2	1,928.0	5.0
Uruguay	6.0	0.2	55.0	1.2	a	a	a	a	a	a
Venezuela	373.0	13.7	993.0	21.0	2,786.0	27.3	2,700.0	16.1	2,175.0	5.6
Other S.A.	—	—	—	—	186.0	1.8	868.0	16.8	729.0	2.0
2. *Central America*	569.0	20.9	727.0	15.3	2,282.0	22.4	4,121.0	24.5	11,675.0	30.0
Mexico	286.0	10.5	415.0	8.8	1,034.0	10.1	2,025.0	12.1	6,962.0	17.9
Costa Rica	30.0	1.1	60.0	1.3	b	b	b	b	b	b
El Salvador	15.0	0.6	17.0	0.3	b	b	b	b	b	b
Guatemala	87.0	3.2	106.0	2.2	b	b	b	b	b	b
Honduras	37.0	1.4	62.0	1.3	b	b	b	b	b	b
Nicaragua	4.0	0.1	9.0	0.2	b	b	b		b	v
Panama	110.0	4.0	58.0	1.2	659.0	6.5	1,458.0	8.7	3,671.0	9.4
Other C.A.	—	—	—	—	589.0	5.8	638.0	3.8	1,042.0	2.7
3. *Other L.A.*	611.0	22.4	761.0	16.1	b	b	b	b	b	b
Cuba	526.0	19.3	642.0	13.6	—	—	—		—	—
Dominican Republic	71.0	2.6	106.0	2.2	b	b	b	b	a	a
Haiti	14.0	0.5	13.0	0.3	b	b	b	b	a	a
4. *Other W. Hemisphere*	n.a.	n.a.	290.0	6.1	1,311.0	12.8	3,130.0	18.6	9,099	23.4
Bahamas									2,987.0	7.7
Bermuda									10,353.0	26.6
Netherlands Antilles									-6,664.0	-17.1
Trinidad & Tobago									932.0	2.4
Others									1,491.0	3.8

Source: Ramsaran 1985, Table 4.3.

investments continued to grow, and hence the US share of the total FDI in the region grew as well. In terms of industry sectors, oil production in Venezuela and elsewhere (especially Mexico and Ecuador) grew most rapidly, helping to bring FDI in Latin America up to its highest point since the turn of the century, almost 40 per cent of total US direct investment abroad. The public utilities sector suffered the largest decline in FDI share, as a result of nationalizations throughout the region in the late 1940s (especially government takeovers of properties owned by American & Foreign Power Company).[20]

The 1950s were characterized by short bursts of FDI growth and decline, punctuated by a 340 percent increase in the flow of US FDI in 1956 (primarily new petroleum investment in Venezuela) and a 42 percent decline in 1960 (due to Cuban nationalization of US businesses after the 1959 Castro revolution). The rapid jump in Venezuelan oil investment resulted from the Suez crisis, which led to decreased confidence in the security of Middle East oil supplies and greater demand for sources closer to the US market. Raw materials investment dominated overall in terms of value, though manufacturing investment grew in dollar and in percentage terms. New mining ventures in the early 1950s in Chile, Venezuela, and elsewhere added substantially to the value of US FDI. Manufacturing FDI was comprised mainly of market-serving investments in machinery, chemicals, transportation equipment, and processed foods, and it was largest in the major markets, Brazil and Mexico.[21]

During the 1960s, US direct investment turned more and more to the manufacturing sectors and away from raw materials ventures. By this decade, most of the public utilities had been bought by national firms or governments, mining ventures were being sold to local joint-venture partners or other local investors, and agricultural investments were being sold to local buyers (with the US firms maintaining long-term purchase agreements and other contractual ties.) The increase in manufacturing investment was due both to the desirability of serving growing markets from within and the protectionist government policies in Latin America that penalized imports. In 1966 the US FDI position in Latin American countries as a whole showed $2,973 million in manufacturing industries, $2,188 million in petroleum, $1,066 million in mining, and $526 million in public utilities.[22]

By the late 1960s it was clear that US investors occupied positions of great industrial power in most Latin American countries. Simultaneously, the countries had swung to highly nationalistic outlooks in general, and anti-American views in particular. Many challenges were raised to US companies' domination of important Latin American business sectors, and some highly visible expropriations even took place (e.g. the copper companies in Chile and Peru; Exxon in Peru; ITT in Chile). Government policies shifted to limit foreign investors' abilities

23

to operate in many countries (such as in the Andean Pact) and import substitution became the broad goal of economic policies in many nations of the region.

During the 1970s the nationalistic trend continued, with extractive and utility investors being forced to sell ownership to nationals throughout Latin America. Venezuela nationalized the foreign oil companies in 1975. Peru and Chile continued their policies of nationalization in mining and utilities. However, in industries where foreign investors possessed some technological or market-access capabilities that governments valued highly and could not obtain otherwise, FDI continued to grow. By 1980 the book value of US manufacturing investment in the region dwarfed all other sectors, at $14.5 billion. In contrast, utilities and agricultural investments were so small that they are no longer reported separately, and mining investment had grown only to $1.41 billion. Petroleum investment remained quite substantial at $4.34 billion.[23]

This is approximately the situation that existed in 1982, at the beginning of the Latin American debt crisis that has drastically cut the inflow of direct investment and other foreign capital into the region during the decade of the 1980s. Sectoral shares in total FDI remain similar to those reported above, and the value of US FDI in the region has grown very slowly, as discussed below in Chapter 3. Table 1.9 gives a detailed look at the country and sectoral distribution of US direct investment in Latin America in 1980.

Table 1.9 US FDI stock in selected Latin American countries in selected industries, 1980 (US$m)

Country	Total	Mining	Petro- leum	Total mfg	Chem.	Elec. equip.	Food	Trade	Banking & finance
Argentina	2,494	s	395	1,584	416	43	168	1,432	1,816
Brazil	7,903	141	365	5,145	1,936	375	430	571	1,276
Chile	536	209	91	s	23	s	s	64	33
Colombia	1,012	s	265	548	184	46	76	97	35
Ecuador	322	0	160	113	17	19	27	32	s
Mexico	5,989	95	159	4,489	1,061	405	366	727	195
Panama	3,171	0	503	262	151	1	s	601	1,376
Peru	1,655	s	s	s	33	13	7	64	9
Venezuela	1,908	0	40	1,032	346	39	221	336	160
Total Latin Amer.	38,882	1,625	4,331	14,550	3,594	1,009	1,693	3,872	12,475

Source: US Department of Commerce, *Survey of Current Business*, August 1982.
Note: s data suppressed to avoid disclosing information about individual companies.

Conclusions

FDI from other source countries

Direct investment and other activities of multinational firms historically have been a primarily American and British phenomenon in Latin America. This characterization is changing in the 1980s, as greater manufacturing investment enters from Japan, Germany, and other European countries (as well as some from other Latin American countries). A perspective on the distribution of FDI flows during the 1970s is given in Table 1.10. Clearly, Japanese and German direct investments have far surpassed those from other European countries, though still trailing far behind FDI by American firms.

Table 1.10 Total flow of direct private investment (net) by selected countries to Latin America in the period 1969–78 (US$m)

Country of origin	Argentina	Brazil	Mexico	Andean[a] group	CACM[b]	Other[c] L.A.	Total
W. Germany	93.87	1,069.23	212.94	61.09	9.67	52.08	1,498.88
Belgium	19.98	32.59	4.06	4.61	2.88	34.96	99.08
France	171.45	232.94	27.55	71.7	0.52	29.81	533.97
Italy	47.48	63.49	0.93	12.24	0.77	17.14	142.05
Netherlands	−1.60	114.38	10.89	2.89	−	37.08	163.64
Japan	−5.03	1,776.47	155.64	449.55	36.55	193.63	2,611.84

Source: Ramsaran 1985, Table 4.8
Notes:
[a]Bolivia, Chile, Colombia, Ecuador, Peru, Venezuela (excluding Chile after 1977).
[b]Costa Rica, Guatemala, Honduras, Nicaragua, El Salvador.
[c]Chile (from 1977), Cuba, Dominican Republic, Haiti, Panama, Paraguay, Uruguay.

Major MNEs from around the world are expanding their business in this region: Volkswagen, Toyota, Nissan, the German chemical companies, the Swiss pharmaceutical firms, Japanese electronics companies, and many others have spread their affiliates from Mexico to Argentina. Table 1.11 lists a few of the most active MNEs in Latin America from countries other than the United States. Note that, despite the relative decline of extractive investment in Latin America since the Second World War, the most important MNEs still include a large proportion of oil and mining firms. Also, note that the manufacturing sectors that have attracted most overall FDI in the region are the same ones in which US firms are most active: autos, chemicals (and pharmaceuticals), machinery, and food processing.

Historical and theoretical background

Table 1.11 Largest non-US MNEs in Latin America, 1985

Company	Country of origin	Country of operation	Annual sales in Latin America (US$ m)
Royal Dutch-Shell	UK/Neth.	Brazil	2,673
Volkswagen	West Germany	Brazil	2,033
Royal Dutch-Shell	UK/Neth.	Argentina	916
Pirelli	Italy	Brazil	912
Mercedes Benz	West Germany	Brazil	869
Fiat	Italy	Brazil	853
Nestlé	Switzerland	Brazil	716
Gessy Lever	Netherlands	Brazil	588
Volkswagen	West Germany	Mexico	555
Philips	West Germany	Brazil	499
Mannesmann	West Germany	Brazil	421
Hoechst	West Germany	Brazil	385
Massey Ferguson	United Kingdom	Brazil	369
Renault	France	Argentina	364
Alcan	Canada	Brazil	347
Nissan	Japan	Mexico	329
Ciba-Geigy	Switzerland	Brazil	314
Bayer	West Germany	Brazil	314
Nestlé	Switzerland	Mexico	277
BASF	West Germany	Brazil	264
Saab	Sweden	Brazil	240

Source: *Progreso*, December 1986.

Direct investment relative to other sources of external finance

Since the First World War, portfolio investment and direct investment have been supplemented by several additional forms of long-term capital inflows from the industrial countries. The most important of these initially were government lending and grants, accounting for annual inflows of between $100 million and $200 million during the 1950s. By the 1960s commercial bank lending was playing an important role, which expanded dramatically with the 'recycling' of funds earned by Middle East oil producers in the 1970s. By 1980, commercial banks (mainly from the United States) provided an annual inflow of long-term loans to Latin American borrowers of about $13 billion, almost three times as much as the direct investment in that year. Table 1.12 shows these data and other facts about external financing of Latin America during the post-war period.

26

Table 1.12 Net inflow of external resources and compensatory financing to Latin American/Caribbean region, 1946–81[a]

	Annual averages (US $m)										
	1946–50	1951–5	1956–60	1961–5	1966–70	1971–5	1976–80[b]	1978	1979	1980	1981[c]
Official grants and non-compensatory capital	84	578	1,348	1,116	2,660	8,213	16,956	19,399	20,535	20,507	23,700
Official grants & long-term loans	−204	93	239	438	939	2,636	6,715	7,963	7,587	4,762	5,317
Long-term private capital	331	343	919	683	1,404	6,081	12,466	14,202	13,777	18,717	24,050
Direct investment				396	847	2,003	3,758	3,918	4,993	5,449	6,574
Loans and other items	−6	56	193	287	557	3,078	8,709	10,284	8,784	13,269	17,476
Short-term private capital	−37	87	−2	−5	318	496	−2,226	−2,766	−829	−2,973	−5,666
Compensatory financing	92	177	289	−13	−464	−762	−389	−6,625	−3,289	10,083	21,639
Transactions of the monetary sector	n.a.	n.a.	n.a.	−13	80	1,775	4,364	3,657	3,128	8,014	19,422
Net change in international reserves: (−) increase, (+) decrease	76[d]	45[d]	114[d]	e	543	−2,537	−4,753	−10,299	−6,417	2,069	2,217

Source: Ramsaran 1985, Table 3.4.

Notes:

[a] Figures up to 1960 refer only to the Latin American republics (excluding Cuba in 1959 and 1960). Since 1961 the coverage includes all member countries of the IDB (except Canada, the US and Suriname). Data for Barbados are available since 1964 and for the Bahamas since 1973.

[b] These averages are not strictly comparable with averages for the decade of the 1960s and 1950s. Beginning in 1971, transactions of central banks have been incorporated into the A-1 category. Category B-1 includes only transactions of deposit banks.

[c] IDB estimate.

[d] Only official monetary reserves.

[e] Included under transactions of the monetary sector.

Historical and theoretical background

Government-business relations

A final note is in order here in the context of this historical introduction to the activities of multinational enterprises in Latin America. Far more than in the United States or in many European countries, the government in Latin America plays a major or even dominant role in industry. The number of government-owned firms is large in Latin America. The predominance of these firms in raw materials extraction and public utilities means that they tend to be very large and capital-requiring. Competition with industrial-country firms is highly regulated, and some activities are precluded for foreign firms (especially oil extraction and utility operation). Even in manufacturing industries, the level of government regulation tends to be quite high throughout the region. While it certainly cannot be said that Latin American governments are very similar, none the less the common thread of heavy government participation in the economy does run clearly from Argentina to Mexico.

Relations between host governments and foreign MNEs have run the gamut between very supportive (i.e. offering subsidies) and totally negative (e.g. the expropriations in Chile, Peru, and Cuba). Generally, the underlying goals of the governments are to industrialize and to have local individuals and firms participate as extensively as possible in the economy. These goals are somewhat difficult to reconcile, in that foreign companies and individuals often possess capabilities that locals do not; so restricting business to only locals hinders access to the most modern knowledge and skills. As a result, policy toward foreign MNEs tends to shift from accommodative, when more foreign technology and capital are desired, to restrictive, when economic progress is more satisfactory.

Broadly speaking, three waves of policy have passed during the post-war period. During the years immediately after the Second World War, there was a general desire to attract more foreign investment and foreign technology. Even so, many public utilities and some raw materials ventures were purchased from foreign investors by nationals in several countries of the region. By the mid-1960s, a shift had taken place, and much greater nationalism pervaded Latin America. More nationalizations took place; the Andean Pact countries (and later Mexico) forced foreign investors to sell part ownership to locals; and generally the 'rules of the game' became less favorable to the MNEs. The oil crises reinforced this wave, by giving Mexico, Venezuela, and Ecuador much greater bargaining power in dealing with foreign firms as a result of their oil exports and reserves. When oil prices fell in the early 1980s, and the Latin American debt crisis struck, the latest wave has favored more lenient treatment of foreign MNEs and greater efforts to attract new companies. Indeed, this wave began in the mid-1970s in the non-oil countries, which were hurt by the higher oil prices just like other importing nations in other areas.

More or less simultaneous with these three waves of policy have been economic development strategies that focus on export promotion and import substitution. The export promotion strategy involves greater participation in the international economy and consequently greater willingness to allow foreign firms to enter in export-related industries. This strategy characterized the first and third waves of government policy. The import-substitution strategy favors inward-looking activities, and often promotes local firms at the expense of foreign ones. The 1960s and 1970s were largely characterized by this strategy, and also by relatively unfavorable treatment of foreign MNEs.

These issues of government policy are discussed in more detail in Chapter 4 below. At this introductory stage it simply is useful to establish the time frame for policies and business activities that will reappear below. Next, we turn to the conceptual framework that underlies the rest of the analysis.

Notes

1. Most of the direct investment at that time could actually be categorized as expatriate investment. That is, Spanish and Portuguese entrepreneurs moved to the Latin American countries and set up their own businesses that indeed received financial backing from supporters in the home country. But the new businesses were not typically extensions of firms from Europe; rather, they were new enterprises run by foreign nationals that were financed (partly) from abroad.
2. United Nations, *External Financing in Latin America*, New York: United Nations, 1965 (E/CN.12/649/Rev.1), p. 5.
3. Irving Stone, 'British long-term investment in Latin America, 1865–1913', *Business History Review* (autumn 1968), pp. 313–14.
4. Cleona Lewis, *America's Stake in International Investments*, Washington, DC: Brookings, 1938, pp. 575–607.
5. Stone, p. 311.
6. United Nations, p. 19.
7. The information in this paragraph comes from United Nations, p. 33.
8. United Nations, p. 5; and F.J Rippy, *British Investment in Latin America, 1822–1949*, Minneapolis: University of Minnesota Press, 1959, Ch. 2. Interestingly, the main British mining ventures were located in Mexico and Venezuela, rather than in the Andean countries which attracted so much US mining investment. See Rippy, p. 50.
9. Rippy, p. 33.
10. United Nations, p. 7; and Rippy, Ch. 3.
11. Rippy, p. 41.
12. United Nations, p. 8.
13. Rippy, p. 69; the author notes that 'The most profitable British railway enterprise in Brazil or anywhere else in Latin America was the São Paulo Railway

Company Ltd, organized in 1858 to construct and operate a line from the city of São Paulo down the mountainside to the coffee port of Santos' (p. 154). Investments in other industries included 115 mining ventures, 112 public utility companies, and 77 farming and/or real estate ventures (p. 69).

14. United Nations, p. 32.
15. Mira Wilkins, *The Emergence of Multinational Enterprise*, Cambridge, Mass.: Harvard University Press, 1970, pp. 110, 120.
16. Lewis, pp. 578ff, as quoted in Wilkins, p. 110.
17. Wilkins, p. 162.
18. Mira Wilkins, *The Maturing of Multinational Enterprise*, Cambridge, Mass.: Harvard University Press, 1974, pp. 10-14.
19. Wilkins, *The Maturing of Multinational Enterprise*, p. 104, 133.
20. Ibid., pp. 302-4.
21. United Nations, p. 33; Ramesh Ramsaran *US Investment in Latin America and the Caribbean*, New York: St Martin's Press, 1985, p. 84; Wilkins, *The Maturing of Multinational Enterprise*, Ch. 12.
22. US Department of Commerce, *Selected data on US direct investment abroad, 1966-76*, Washington DC: US Dept of Commerce Bureau of Economic Analysis.
23. Ibid.

Bibliography

Lewis, Cleona (1938) *America's Stake in International Investment*, Washington, DC: Brookings Institution.

Ramsaran, Ramesh (1985) *US Investment in Latin America and the Caribbean* New York: St Martin's Press.

Rippy, J. Fred (1959) *British Investment in Latin America, 1822-1949*. Minneapolis, Minn.: University of Minnesota Press.

Stone, Irving (1968) 'British long-term investment in Latin America,' *Business History Review*, autumn pp. 311-39.

United Nations (1965) *External Financing in Latin America*, New York: United Nations.

Wilkins, Mira (1970) *The Emergence of Multinational Enterprise*, Cambridge, Mass.: Harvard University Press.

—— (1974) *The Maturing of Multinational Enterprise*, Cambridge, Mass.: Harvard University Press.

Chapter two

Theories of multinational business

Introduction

Origins of international business theory

The theory of the multinational enterprise (MNE) has developed rapidly
during the past three decades, as large raw materials, manufacturing,
and service companies with affiliates in several countries have come to
the forefront in international business activity. During the same time
period, international business itself has demonstrated shifts toward
increased foreign direct investment (and the production that it generates)
and toward greater importance of manufactured goods and petroleum
products in that business. The vast majority of this business continues
to take place between industrial countries.

Within Latin America similar trends have surfaced. Trade grew
dramatically during the inflationary 1970s. Direct investment grew also,
but more slowly, due largely to the inward-looking development strategies
followed by Latin American governments during the 1970s. In particular,
natural-resource industries were largely nationalized in the region,
pushing foreign MNEs into positions of technology and management suppliers and buyers of the extracted materials. Foreign direct investment
(FDI) in mining actually fell between 1970 and 1980, while FDI in the
oil business grew only very slightly. Manufacturing and service-sector
FDI, on the other hand, grew rapidly during the decade.

Then in the 1980s the region-wide recession cut off growth in both
trade and FDI. Table 2.1 portrays these trends. Other than exports, every
other category of international business in the region declined during
the recession and debt crisis of the 1980s. Indications are that slow growth
had resumed by 1987, but full recovery from the downturn remains to
be documented. As in the rest of the world, manufacturing has become
the dominant form of FDI, with petroleum remaining in second place.
The region's current account balance of payments shows a recent shift
toward equilibrium away from a huge deficit; this is consistent with

slow growth in the region combined with more rapid increase in demand in the United States and Western Europe. (It also implies the success of a reorientation of government policy toward export promotion and away from inward-looking import substitution.)

Table 2.1 Trends in international business in Latin America and the Caribbean (US$m)

Activity	1950	1960	1970	1980	1985
Exports	6,611	9,264	15,923	103,719	104,932
Imports	5,855	9,622	17,241	113,795	74,238
Current account balance of payments			−3,573	−27,436	−2,707
Foreign direct investment inflow	372*	−521	1,077	5,709	1,953
US FDI *stock* by industry					
— Total	4,735	8,366	12,961	38,275	29,479
— Manufacturing	780	1,521	4,541	14,489	15,323
— Petroleum	1,408	3,122	2,703	4,336	5,299
— Mining & smelting	628	n.a.	1,712	1,408	n.a.
— Other	1,919	3,723	4,005	18,042	n.a.

Sources: International Monetary Fund, *Direction of Trade Yearbook*, 1986; US Department of Commerce, *Survey of Current Business* (various issues).
*Data for 1951.

These data refer to international business and MNEs in Latin America. A theory of international business, and particularly one aiming to explain the activities of MNEs in Latin America, needs to offer reasons for the occurrence of these phenomena and some suggestions of what to expect in the future. In particular, such a theory needs to focus on the microeconomic level, looking at (multinational) firms' decisions and government policies that affect them.

All of the trends in Table 2.1 relate directly to the growth of multinational enterprises. For example, FDI is the necessary condition for a multinational enterprise; that is, to be multinational, a firm must invest in affiliates in at least three countries. Also, manufacturing is the base of far more multinational firms than agriculture or extractive ones. In fact, the diminishing costs of international transportation and communication (along with supportive or at least not overly-restrictive government policies) that enabled MNEs to flourish also brought with it a boom in trade among industrial countries, especially in manufactured goods. Before this period, theorists generally ignored MNEs or subsumed them under the heading of international trade in analyzing international economic or business phenomena.

Probably the major writing that marked a turning-point in international business analysis was Vernon's international product cycle (Vernon 1966). That model offered a formal alternative to Ricardian comparative advantage, as amended importantly by Heckscher, Ohlin, and others. Instead of focusing on production costs that vary by country as the basic issue determining international business transactions, Vernon looked at technological innovations and marketing strategy as key factors that enable firms to successfully compete through foreign direct investment, as well as exporting. International business then evolved as a discipline with greater micro (i.e. company–level) focus and more empirical orientation than its counterpart in economics.

Despite the divergence of paths between business and economic analyses of international phenomena, much of the international business literature is based on economic theory; and economic analysis in recent years has developed a branch that explores the industrial organization of international firms. Thus theories of international (and multinational) business come from both disciplines, and advances in understanding of the phenomena under study often have ramifications in both areas.

Multinational vs. international

This chapter, and the entire book, examine issues related to multinational firms, rather than international business in general. That is, the analysis is concerned with firms that operate in three or more countries and which maintain information, financial, and control links between affiliates. The theory to be pursued similarly relates to the many aspects of such firms rather than to the full spectrum of international business activities that include exports by non-MNEs and international banking by banks that do not operate locally in foreign countries. (Generally, the theory can be applied to these phenomena as well, but the full implications are not discussed here.)

The level of analysis chosen for this book is microeconomic, that is, focused on individual firms and on the issue of government regulation as it affects individual companies. Moreover, the focus is on multinational firms, as defined previously, since they represent the major actors in activities such as foreign direct investment, international licensing and contracting, international banking, and international trade.

By taking this approach, the path is clear to explore intra-company activities such as financial transfers, strategic decision-making, and organizational structures that play important roles in multinational firms. These activities may be seen as the functioning of the MNEs' internal market (or hierarchy), since they represent an alternative to outside markets as a means of organizing economic activity. (See the discussion of markets and hierarchies below on pp. 37 and 38.) In addition, regulatory issues are considered only as they relate to

33

MNEs, not with respect to all aspects of international business.

Economics and business strategy

The two bodies of theory that bring the most to bear on analysis of MNEs are those of economics and business strategy. The former includes a wide variety of approaches, beginning with extensions of classical trade theory, and most recently deriving from transactions cost analysis. On the business strategy side, analyses of the nature of competition have led to fruitful approaches such as the internalization view and the pursuit of competitive advantages. Discussion of several of the specific approaches constitutes the bulk of this chapter.

Relevance to Latin America

Just as less-developed countries are concerned about the 'appropriateness' of technology transferred to them from industrial nations, (namely because it tends to be more capital-intensive than local conditions warrant), analysts of business in Latin America need to be concerned about the appropriateness of models and theories created to explain phenomena in the industrial countries. For the most part, theories of multinational business do include less-developed nations in their scope of application. However, there is a notable lack of emphasis on government–business relations in the mainstream of theory, which perhaps is a significant failing, given the great importance of governments in determining and often changing the rules of the game for business in this region. Although a great deal of game theory has been developed in economics to describe government–business relationships, this largely has not been applied in the international business context. The idea of the 'obsolescing bargain,' in which government bargaining power rises over time in dealing with an MNE as the MNE commits more and more of its resources to the host country (i.e. the MNE's power obsolesces) is not integrated into the main body of theory. These differences cannot be handled simply with amendments to the existing theories. The approach of this book attempts to capture some of the government influence on the 'game' between governments and MNEs.

The final section of this chapter is devoted to sketching the conceptual view used throughout the book. The term 'bargaining theory of the MNE' is used to describe MNE activities in Latin America, from competition with rivals to dealing with host governments. Bargaining theory considers MNEs as actors in a multi-player game, in which governments set rules and seek to channel company behaviour into desired activities, while companies attempt to outcompete their rivals in each market and to minimize the burdens imposed by governments. A logical, consistent framework is presented and then explored in the rest of the empirical chapters. Before entering into that discussion, let us consider the major theories that deal with international business and note their relevance to MNE activities in Latin America.

Economic theories of multinational business

Comparative advantage

Classical international trade theory is based on the concept of comparative advantage. This theory was first presented in detailed terms by David Ricardo in 1807, and it has survived almost two centuries of changing economic environments, while continuing to offer valuable insights and clear analysis. Without repeating what can be found in any introductory textbook in international economics or international business (e.g. Lindert 1986), a brief review of comparative advantage is in order here. The comparative advantage theory explains how differing production cost conditions in different countries lead to patterns of trade between them. A country that can produce some products more cheaply than another country will profitably export those products and import ones that are more efficiently produced by the second country, *ceteris paribus*. In this situation both nations will gain in economic welfare, because both will gain from the other's relative efficiency and from the exchange of efficiently produced products.

Comparative advantage theory is used at the macroeconomic level of analysis, that is for analyzing inter-country patterns of trade rather than company export or import dealings. Conclusions about what individual firms do or should do are not derivable from the theory, nor are insights into non-cost factors such as marketing strategy. In fact, comparative advantage usually is applied to production costs only, ignoring the costs of transportation, obtaining information, and many others. Recent authors have extended the comparative cost view to include other costs (e.g. Hirsch 1976) and even demand conditions (e.g. Grosse 1985). Both of these authors and several others move the level of analysis to the firm, and then they consider national cost characteristics as the basis for locating production and exporting.

A second principle of international trade is based on natural resource availability. Clearly, a nation that possesses resources such as petroleum, copper, fertile agricultural land, and so on, can produce and export those raw materials — and nations that do not must import them. While this statement may note an obvious point, nonetheless, resource availability is the basis for a major part of total international trade.

Resource availability and comparative advantage together explain a large part of trade flows between countries. In fact, it could be argued that comparative advantage has been reinvigorated by the post-Second World War phenomenon of offshore assembly, in which firms send raw materials and components for manufacturing some product (such as textiles or electronics) overseas to a low-cost production site for assembly, and then ship the final products to the market(s) that the firm wants to serve.[1] This cost-based strategy gives new importance to the comparative advantage theory, which itself had been greatly criticized for its inability to explain trade in manufactured goods between industrial countries during the past several decades. Indeed, the international product cycle was the most successful effort to come to grips with the empirical failings of comparative advantage after the Second World War.

International product cycle

Vernon's international product cycle marks an effort both to explain the phenomenon of trade in manufactured goods and to shift the level of analysis to the firm instead of the country. This theory uses two basic ideas to reorient thinking about international trade. First, as widely recognized at the time (1966), technology was and is a crucial factor of production. Creation of new technology enables individual companies (and their countries of origin) to be able to produce more efficiently than other firms, or to introduce new products that simply are not available elsewhere. These technological advantages themselves contribute greatly to the patterns of trade in manufactured goods today. The second basic idea was to focus on marketing as another key to international business activity. Companies which could successfully convince customers in the home country to buy a new product, subsequently could look to overseas markets and become exporters. At the macroeconomic level, countries which offer the best systems of distribution and promotion are the most likely candidates for introduction of new products; so industrial nations such as the United States are expected to be the sources of the bulk of new products.

The international product cycle takes one more step beyond traditional trade theory, and becomes the first comprehensive theory of international business. Vernon reasons that, once successful innovations have taken place in one country, the initiating firm may export to other large, relatively high-income markets, *and* the firm may subsequently choose to undertake foreign direct investment to produce locally in the market country. Thus, both exports and foreign direct investment are treated in this unified theory of international business. By focusing at the level of the firm, this theory is able to capture company strategies that cross national boundaries and which do not fit well into the comparative advantage model.

Monopolistic competition

Once the stage had been set to look at international trade and direct investment from the viewpoint of the firm (namely in the international product cycle), theorists began to pursue additional characteristics of MNEs that enable them to compete successfully as exporters, investors, and in other kinds of business. This shift moved theories into the realm of industrial organization, or the institutional side of microeconomics. This body of literature has long focused on issues such as barriers to market entry in different domestic industries, and sources of monopoly rents (i.e. competitive advantages) that some firms possess.

Caves (e.g. 1971, 1974, 1982) has been one of the leading industrial organization theorists to pursue MNE analysis. He explored the importance of marketing advantages as well as technology in US-Canadian and US-European international business during the early 1970s, and has written extensively about other countries and other aspects of monopolistic competition among international firms. The emphasis in Caves' studies of international business generally has been on explaining what attributes of particular firms have led them to invest abroad or export, in comparison with local firms abroad and other MNEs.

Transactions costs

Transactions cost economics gained a great following in the mid-1970s, particularly through the work of Williamson (1975, 1981). Subsequently, Teece (1977, 1986) and many others have applied the basic ideas to multinational firms and international business. This body of literature explores the organization of economic activity between markets and non-market institutions such as governments and oligopolistic firms (called 'hierarchies'). By emphasizing the empirical facts that firms and individuals possess imperfect information, and that they are likely to strategize to maximize their own benefits from business transactions (i.e. they demonstrate 'opportunistic behavior'), Williamson shows that in many situations organizational hierarchies such as MNEs may be more efficient than the free market. Thus the goal for the MNE is to discover those situations where it does possess superior efficiency relative to smaller, 'perfectly competitive' firms, and to employ its resources in such businesses.

Transactions cost theory is quite useful in analyzing MNEs, since such firms clearly constitute hierarchies, with their own internal markets (hence the appropriateness of internalization as a concept in this context). In fact, MNE analysis during the 1980s has followed conceptual bases related to transactions costs more than any other theory. Perhaps the major drawback of this theory is the relative lack of empirical work

37

that has utilized it, due largely to the difficulty in specifying the transactions costs involved. As in many other areas of research, the principle is clear but the measurement lags far behind.

It should be noted that there are two sides to the hierarchy issue. One is the efficiency aspect, as already discussed. The other is the market-power aspect; the large, oligopolistic firm gains some ability to influence prices, quantities, and other features of the market due to its size and its superior information relative to purely competitive firms. Thus, from a regulatory standpoint, both the efficiency and the power of the MNE must be considered in any evaluation of the firm.

Dunning's theory

Dunning (1977) introduced an 'eclectic theory of international production' that seeks to combine the concepts of monopolistic competition and internalization (discussed in the next section). He reasons that three sets of factors form the basis of MNE activity: ownership, location, and internalization advantages. Ownership advantages arise from company- or product-specific characteristics such as proprietary technology, an international distribution network, economies of scale in production, marketing, and financing, and so on. Location advantages arise from the firm's possession of facilities in low-cost locations, or locations with high market potential, or with relatively favorable government regulation, and so on. These are primarily country-specific factors. Finally, MNEs benefit from internalizing those business activities that they can carry out internally more profitably than by dealing with other firms or customers in the market. Together, these three sets of factors constitute major bases of competitive strength for MNEs in comparison with their local competitors and other MNEs in any market.

Dunning's theory is quite useful in that it provides a set of criteria for investigating MNEs (namely the ownership and location advantages) and a mechanism through which MNEs make decisions (namely internalization). This theory is largely an enumerative classification system for characteristics that enable MNEs to exist and compete successfully, rather than a logical, abstract model such as comparative advantage or the international product cycle. None the less, it provides many valuable insights into the operation of MNEs.

Business strategy theories of multinational business

Internalization

It is debatable whether this concept should be discussed under the heading of economics or of business strategy. Internalization relates to the study of corporate vertical and horizontal integration. The focus of internalization is on the firm's decisions whether to integrate upstream or downstream business activities (or similar business in other locations) into its own hierarchy — this contrasts with the economics literature on such integration, which focuses primarily on industry structure and implications of that structure for public policy.

Internalization is a powerful concept, since it provides a rationale for company expansion into new markets and products — a style of growth that typifies large international companies today. Its simplicity also is appealing: the firm should internalize additional activities up to the point that the marginal gain from internalizing is outweighed by the marginal cost of doing so. This reasoning fits the conglomerate growth of ITT in the 1970s (when that firm expanded from telecommunications into hotels, food products, car rental, insurance and other financial services, industrial machinery, and other products through a network of subsidiaries around the world) as well as the comparatively narrow set of product lines of Nestlé (the food products multinational that also operates subsidiaries around the world). Perhaps the greatest difficulty in using the concept of internalization is its high level of generality; it needs additional content such as Dunning's location and ownership advantages or more simply the competitive advantages framework of Porter or Grosse in order to produce clear implications for strategy and policy. Internalization is a concept that explains the growth of the firm, rather than specifically explaining activities of multinational firms.

Competitive advantage

The competitive advantage theory can be viewed as adding more concrete content to the internalization principle. While the idea of competitive advantages is not at all new, this view has gained a wide following during the 1980s. Probably the most detailed presentation of the theory appears in Porter (1985), although the ideas previously were developed elsewhere in a variety of sources (e.g. Spence 1977). The presentation here follows Grosse (1985a).

Assuming a Williamsonian world of imperfect information ('information impactedness and bounded rationality') and opportunistic behavior, this theory explores the kinds of strengths and weaknesses that a firm may possess or obtain in order to gain a preferred (monopolistic) position for itself relative to competitors in the same market. These capabilities include proprietary technology and marketing skills — both of which appear prominently in the other literature discussed previously — as well as skill in managing people, scale economies of various types, superior location, and preferential government relations. In each case, the competitive advantage gives the possessor firm a lower cost, greater revenue, or lower risk than its competitors. Some of the major advantages are listed in Table 2.2

Table 2.2 Selected competitive advantages of multinational enterprises

Advantage	Description	Example of product or company
General Competitive Advantages		
1. Proprietary technology	Product or process technology held by a firm that others can obtain only thru R&D or contracting with the posesssor	IBM; DuPont; Johnson & Johnson; Sony
2. Scale economies in production	Large-scale production facilities that lower unit costs of production	US Steel; Ford; Boeing
3. Scale economies in purchasing	Lower costs of inputs through purchasing large quantities	Any very large firm
4. Scale economies in financing	Access to funds at a lower cost for larger firms	AT & T; IBM; EXXON
5. Scale economies in distribution	Sales operations in several countries, allowing a firm to serve a 'portfolio' of markets	Any very large firm
6. Scale economies in advertising	Sales in several countries allowing somewhat standardized advertising	Coca-Cola; IBM; Nestlé
7. Goodwill based on brand or trade name	Reputation for quality, service, etc., developed thru experience	RCA; Levis; Sears; Hoover; Caterpillar
8. Government protection	Free or preferential access to a market limited by government fiat	General Dynamics; Renault; BP
9. Human resource management	Skill at fostering teamwork among employees & optimizing their productivity	Large Japanese MNEs; IBM
International Competitive Advantages		
10. Multinational market access	Knowledge of and access to markets in several countries	Any MNE
11. Multinational sourcing capability	Reliable access to raw materials, intermediate goods, etc., that reduce single-source risks	Any MNE
12. Multinational diversification	Operations in several countries so country risk and business risk are reduced	Any MNE
13. Managerial experience in several countries	Skill for managing multi-country operations gained thru experience in different countries	MNEs with managers experienced in int'l business

Each of these competitive advantages enables the firm to establish some degree of monopoly power relative to its rivals, and is sustainable for a shorter or longer time period. The strategic goal is to optimize the firm's combination of advantages, such that a profitable, sustainable business position can be carved out. Given the ability of rival firms to copy competitive advantages, the theory emphasizes methods of making advantages sustainable and also the need to create new advantages over time, as old ones are competed away.

The application of this theory to multinational firms focuses on such firms' abilities to utilize basic competitive advantages across national borders. (For example, technology can be shared among affiliates without being used up; economies of scale often can be realized through multi-country sales or purchases, when they are not achievable in one national market.) Also, several important advantages require multinationality for their existence. The bottom part of Table 2.2 notes the key multinational advantages, which include the portfolio effects of operating in multiple markets and supply areas, as well as the managerial experience that may be applicable from one country to another. Figure 2.1 suggests some of the means available to international firms for taking advantage of their competitive advantages.

The theory of competitive advantage posits that advantages such as those listed above exist in every business. The firm's strategy should be to define the relevant advantages, decide which ones to acquire and protect, and create a method of implementing them in the business. The theory is dynamic in that such strategy must be redefined regularly as demand and cost conditions change, as rival firms alter their strategies, as government regulation changes, and as the firm's own internal conditions shift over time.

Clearly this theory is oriented toward company decision-making. However, it still can be utilized in government policy-making at the macroeconomic level. That is, if government policy-makers understand the motivations of company managers, as well as the firms' strengths and weaknesses, they (i.e. governments) will be able to define policies that guide business into the activities desired by the government. For example, by understanding that proprietary technology is the key advantage for many firms in a variety of industries, host governments can establish policies for protection of intellectual property (namely technology) that attract MNEs either to carry out research and development locally, or at least to employ their best technology locally without fear of loss of protection. As another example, if goverments realize that MNEs often have multiple site choices for production or other activities, then they will select regulatory policies that do not overburden MNEs relative to policies in alternative countries. Each of these two issues has been the source of substantial government–MNE conflict,

Table 2.4 Methods for exploiting competitive advantages through foreign involvement

Form of foreign involvement	Added income	Capital commitment	Management commitment	Technology commitment	Political risk	Flexibility	Impact on rivals
			DECISION CRITERIA				
EXPORTS	?	low	low	low	low	high	?
EXAMPLES:	(a) direct						
	(b) through a distributor						
	(c) through a trading company						
CONTRACTING	?	low	possibly high	possibly high	low	?	?
EXAMPLES:	(a) licensing technology						
	(b) franchising						
	(c) management contracting						
PARTIALLY-OWNED DIRECT INVESTMENT	?	?	possibly	high	med.	low	?
EXAMPLES:	(a) joint venture with local company						
	(b) joint venture with foreign company						
	(c) joint venture with government						
WHOLLY-OWNED DIRECT INVESTMENT	?	high	high	low	high	?	?
EXAMPLES:	(a) assembly plant for local sales						
	(b) basic manufacturing						
	(c) raw materials' extraction						
	(d) offshore assembly plant						

Note: Each column represents a dimension for decision-making that should be considered when choosing a method to use in exploiting a competitive advantage. The rankings will differ from company to company and also across countries; the entries in the table are for illustrative purposes only.

when by using alternative policies the government could have avoided the negative implications (e.g. the exodus of the MNE(s) involved). Thus this micro-level theory has implications for macro-level policy-making.

This book's approach: a bargaining theory of the MNE

So much of a foreign firm's success in Latin America depends on dealing with the host government that it is almost natural to turn to a theory of government–business relations as the conceptual basis for understanding company activities and performance. Despite the fact that virtually all theories of the firm (multinational or otherwise) concentrate on the characteristics of the firm, and sometimes on the relevant aspects of rival firms, the context of competition in Latin America calls for a different view. The main reason for this difference may be the level of economic development in the region compared to industrial countries where most firms operate and theories arise, or it may be due to the statist tradition in Latin America, or some other factor such as industrial concentration in these markets. The result is that company activity is heavily influenced by government policies, which change relatively rapidly and significantly compared to policies in the industrial countries.

The bargaining theory advanced here is an extension of the competitive advantage approach outlined above. In addition to competition with rival firms, the bargaining theory posits competition with governments for the distribution of those revenues and expenses (and externalities) generated by the firm. Since the government in each country establishes the rules of the game for MNEs operating there, competition with governments is generically different from that with rival firms. That is, governments not only compete for economic gains but they also define the rules of competition. Since the governments and companies are players in a game of resource allocation, and their success is dependent on relative strengths and weaknesses, the situation in this sense is quite similar to competitive strategy between firms.

To clarify the competitive nature of the relationship between MNEs and governments, consider the position of a potential foreign direct investor and the government of a given country. Table 2.3 presents some major, broad categories of bargaining advantage that often exist for MNEs and for governments. Note that each player controls one or more factors that are of basic necessity to the other. Namely, the government controls access to a country's market and/or to its production inputs; while the MNE controls a source of technology and access to foreign markets, each of which typically could be obtained by the country only at a (much) higher cost through other means. The game involves positioning and decision-making by both company and government in an ongoing bargaining relationship, with the goal of each player being to maximize its own gains without causing the other to withdraw from the game.

Table 2.4 Bargaining advantages of MNEs and governments

Governments	Firms
1. Sovereignty over raw materials	1. Control over technology
2. Control over market access to the domestic market	2. Control over international transmission of information
3. Right to establish and change the 'rules of the game' on all business	3. Ability to move production from one country to another
4. Right to establish and change the rules on MNEs, such as taxes, permissions, subsidies, ownership	4. Flexibility to change strategy and activities, limited only by preferences of owners
5. Ability to play MNEs against each other to obtain more benefits	5. Market access to other countries through affiliates

The MNE would like to invest as few financial, physical, and human resources as possible and obtain the largest return possible for its shareholders. Optimally, the MNE would like to produce in a small number of very low-cost locations and ship its products or services to customers around the world where markets justify. In relatively small markets (such as most Latin American countries) the MNE would often prefer to export from an existing production facility elsewhere to serve the small market, unless constrained by the government from importing.

The host government, on the other hand, would like to maximize the investment made by MNEs, and to channel their earnings back into the local economy. The government primarily seeks stimulation of the economy, which the MNEs can do through their investment, consumption of local inputs, and exporting to other countries. In addition, governments typically seek more employment, assistance in improving the balance of payments, and inflow of modern industrial technology. Governments do not want MNE control over their economies, so the issue of sovereignty precludes them from opening the regulatory door too wide.

Obviously, while there are some congruent goals (such as good economic performance by MNEs in the host countries), there are quite a few conflictive goals between companies and governments. A basic task of the bargaining theory is to demonstrate the likely outcomes of conflictive situations and to explain what the results would obtain. This task is undertaken empirically in Chapter 4 below.

The bargaining relationship generally implies a diminishing or obsolescing position for the MNE, which loses some of its power as it commits more and more resources to facilities in the host country, as it trains local people in operation and management of those facilities, and as the government learns how to deal with the firm over time. The *obsolescing bargain* is one of the reasons that host countries among the LDCs have successfully induced MNEs over the years to accept local joint-venture partners, to reinvest most of their local earnings, and to employ virtually all local nationals in their facilities.

A set of clear examples of the obsolescing bargain in Latin America are the national oil industries of countries such as Venezuela and Ecuador, and the copper industries in Chile and Peru. In each of these cases, the industry was initiated many decades ago by foreign MNEs that explored for the natural resource, extracted it, and used or sold it in other (typically industrial) countries. By the 1960s, local managers and engineers, many of whom were trained abroad, had learned most of the technology necessary to operate the extractive operations. In addition, the government had learned a great deal about dealing with the MNEs over decades of negotiating and regulating. When nationalistic movements swept Latin American countries in the late 1960s, these two industries were widely nationalized, with ownership and control going to local nationals, and technical service contracts being offered to the foreign MNEs. In Venezuela, the oil industry actually was not nationalized until 1975, at about the end of the inward-looking period in the region. The obsolescing bargain caught the extractive companies such as EXXON, Texaco, Shell, Anaconda, Kennecot, and Cerro de Pasco with large-scale facilities exposed in the host countries and relatively little bargaining power to preclude more onerous government regulation.

The other side of this coin is that MNEs have not seen their bargaining positions dwindle away completely. Development of new technology, access to foreign markets, and other bargaining advantages remain for the firms; so the game continues without resulting in the demise of the MNEs. Both technology and foreign-market access have proven to be sustainable bargaining chips for the multinationals in their dealings with home and host governments around the world.

As a contrast to the extractive industries, numerous examples can be called on in high-tech industries in which foreign MNEs have remained relatively powerful in their long-term bargaining strength relative to governments. The computer industries in every country except Brazil continue to be dominated by IBM, Burroughs, Olivetti, and a handful of foreign MNEs.[2] Aircraft manufacturing remains almost exclusively the domain of foreign producers, with the exception of Brazil's Embraer. Pharmaceuticals companies have been pushed into producing locally throughout the region, but with very little local competition or local ownership. And finally, export-oriented firms face few ownership restrictions in Latin American countries, which are favorably affected by the balance-of-payments impact of such firms.[3]

The nature of any particular government–company bargain can be viewed as dependent on the relative importance of the situation to each actor, and their bargaining advantages relevant to the particular situation. Figure 2.2. gives some idea of the criteria that determine bargaining strategies of both the firms and the governments. Note that each side (company and government) will pursue more aggressive or more cooperative strategies depending on what is at stake, how much power it has in the given situation, and what the existing relationship is between the two sides at the time.

Figure 2.1 Characteristics of the government–company bargain

Key relationship dimensions:

1. Stakes of the firm and of the government in the conflictive issue.
2. Power of the firm relative to power of the government in the conflict.
3. Interest interdependence: is the conflict zero-sum or not?
4. Quality of the relationship between the two sides.

Two-dimensional representations of govt–business relations

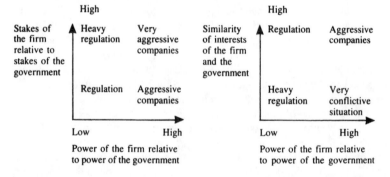

A more concrete perspective is in order. The bargaining theory hypothesizes that, in any situation such as the transfer of technology, the government's and company's power positions will lead to some outcome that favors one relative to the other. For example, several of the countries in Latin America have forbidden MNEs to pay royalties for the use of their technology from a subsidiary back to the parent firm. Apparently, these countries view their power as sufficient to force company behavior without driving away the desired investment, employment, and so on that MNEs bring. The success or failure of this policy can be measured empirically. As another example, the Andean Pact countries and Mexico have tried to force MNEs to accept local joint owners and local control of their facilities by demanding local majority ownership of companies. The results of this policy can be measured to see how successful the governments have been. Similarly, many other bargaining situations can be evaluated with respect to each party's power, each party's stakes, their interest interdependence, and the empirical results. The discussion in this section has been descriptive and suggestive, rather than empirically based. Selected issues that illustrate the bargaining theory are examined in Chapter 4. Some simple tests of the bargaining theory also are laid out there.

The intent in presenting this bargaining theory of the MNE is to raise issues that have great empirical importance in Latin America, and to demonstrate some of the key motivating factors that lead both companies and governments into the policies that they pursue. Issues such as joint-venture requirements (in Mexico and the Andean Pact), profit remittance limits (in the Andean Pact and Brazil), technology constraints (throughout the region), and protection policies (also throughout the region) can be seen as results of the bargaining process. Thus the bargaining theory forms the framework for the analysis that appears in the following chapters.

Notes

1. See, for example, Sanjaya Lall 'Offshore assembly in developing countries', National Westminster Bank *Quarterly Review* (August 1980).
2. The computer industry provides an interesting contrast with respect to the obsolescing bargain. In Mexico during 1985, IBM was given permission to build a 100 percent foreign-owned plant to manufacture computers; while in Brazil also in 1985, all foreign computer companies were denied permission to import *or* produce locally micro- and mini-computers, so that the market would be reserved to local firms. In the former case the company's bargaining position has not been eroded, but in the latter the government has challenged the foreign computer firms with a policy that may (or may not) produce competitive Brazilian computer makers and exclude foreign MNEs.
3. Kobrin (1987) found that manufacturing firms tended *not* to experience an obsolescing bargain in LDCs, contrasting with the experiences of extractive firms.

Bibliography

Caves, Richard (1971) 'International corporations: the industrial economics of foreign investment' *Economica* (February).
(1974) 'The causes of direct investment: foreign firms' shares in Canadian and UK manufacturing industries' *Review of Economics and Statistics* (May).
(1982) *Multinational Enterprise and Economic Analysis*, Cambridge: Cambridge University Press.
Dunning, John (1977) 'Trade, location of economic activity, and the MNE: a search for an eclectic approach,' in Bertil Ohlin (ed.) *The International Allocation of Economic Activity*, New York: Holmes & Meier.
Gladwin, Thomas, and Ingo Walter (1980) *Multinationals under Fire*, New York: Wiley.

Grosse, Robert (1981) 'The theorys of foreign direct investment,' University of South Carolina *Essays in International Business* (December).

(1984) 'Competitive advantages and multinational enterprises.' University of Miami *Discussion Papers in International Business* (December).

(1985) 'An imperfect competition theory of the MNE', *Journal of International Business Studies* (January).

Hennart, Jean-François (1982) *A Theory of Multinational Enterprise*, Ann Arbor, Mich.: University of Michigan Press.

Hirsch, Seev (1976) 'An international trade and investment theory of the firm,' *Oxford Economic Papers* (July).

Hout, Thomas, Michael Porter, and Eileen Rudden (1982) 'How global companies win out,' *Harvard Business Review* (September–October).

Johansson, Johny (1983) 'Firm-specific advantages and international marketing strategy,' Dalhousie *Discussion Paper No. 24* (May).

Kobrin, Steven (1987), 'Testing the bargaining hypothesis in the manufacturing sector in developing countries', *International Organization* (Autumn).

Lecraw, Donald, (1984), 'Bargaining power, ownership, and profitability of transnational corporations in developing countries', *Journal of International Business Studies* (Spring/Summer).

Lindert, Peter (1986) *International Economics*, 8th edn, Homewood, Ill.: Richard D. Irwin.

Lindert, Peter, and Charles Kindleberger (1982) *International Economics*, 7th edn, Homewood, Ill.: Richard D. Irwin.

Newman, Howard (1978) 'Strategic groups and the structure–performance relationship,' *Review of Economics and Statistics* (August).

Ouchi, William (1979) *Theory Z*, New York: Basic Books.

Porter, Michael (1980) *Competitive Strategy*, New York: Free Press.

(1985) *Competitive Advantages*, New York: Free Press.

Ricardo, David (1807) *Principles of Political Economy and Taxation*, Sussex, Penguin. (1971, Harmondsworth: Penguin.)

Rugman, Alan (1981) *Inside the Multinationals*, London: Croom Helm, 1981.

Shapiro, Daniel (1983) 'Entry, exit, and the theory of the multinational corporation,' in Charles Kindleberger and David Audretsch (eds) *The Multinational Corporation in the 1980s* Cambridge Mass.: MIT Press.

Spence, A.M. (1977) 'Entry, capacity, investment, and oligopolistic pricing,' *Bell Journal of Economics* (autumn).

Teece, David (1977) 'Technology transfer by multinational firms: the resource cost of transferring technological knowhow,' *Economic Journal* (June).

(1986), 'Transaction cost economics and the multinational enterprise', *Journal of Economic Behavior and Organization*.

Vernon, Raymond (1966) 'International investment and international trade in the product cycle,' *Quarterly Journal of Economics*.

Williamson, Oliver (1975) *Markets and Hierarchies: Analysis and Antitrust Implications*, New York: Free Press.

(1981) 'The modern corporation,' *Journal of Economic Literature* (December).

Chapter three

Recent trends in multinational business in Latin America

Introduction

A label that characterizes most of multinational business in the Latin American region during the 1980s is retrenchment. Due substantially to the debt crisis (discussed in some detail in Ch. 7), growth in exports, direct investment, bank lending, and other forms of international business have fallen to low or even negative rates for most of the decade. Both industrial and financial MNEs have generally tried to limit their commitments of resources to Latin American countries since 1982, when the debt crisis began with the Mexican government's declaration of an inability to service its foreign official debt. In the case of commercial banks and other financial institutions, this has meant an attempt to reduce the value of loans to Latin American borrowers. In the case of industrial companies, this has meant an attempt to avoid additional investment in the region and to remit as much of available funds as possible to parent companies. These tendencies have followed the overall economic stagnation of the region, with per capita GDP falling substantially in most countries and in 1987 only approaching that of the mid-1970s in real terms.

To put this situation into perspective, it should be noted that the Asian less-developed countries (including China) surpassed Latin America as the principal LDC host region for foreign direct investment during the first half of the 1980s.[1] The Asian nations are attracting FDI not only to serve local markets, but also very importantly to take advantage of low-cost manufacturing sites for products that are ultimately sold in industrial-country markets. Although this last phenomenon does take place in Latin America, it is generally on a very small scale, with the very major exception of the 'maquiladora' zones on the Mexican borders with Texas and California. In addition, the Asian LDCs were not as successful in attracting foreign bank loans during the 1970s, so they have faced a much less severe debt crisis during the 1980s. While it is not the purpose here to compare these two regions, Latin America's crisis

49

(and its possible means of resolution) should be understood as particular to that region rather than symptomatic of less developed countries in general.

Returning to the variety of international business activities in Latin America, in value terms the amount of foreign bank debt dwarfs all other types of business. This fact is especially problematic, since the bank debt is that part of international business which has been most severely limited by the crisis. The drastic reduction in new funds' availability to Latin American borrowers from outside lenders has contributed to the decline in capital formation in the region. For the first time since the Depression of the 1930s, North American and European capital markets have been virtually closed to Latin American borrowers.

Aside from the (hopefully) transitory problem of foreign debt, one of the main determinants of MNE activities in all Latin American countries is government policy. In all countries, sectors such as some raw materials and public utilities are restricted to domestic, often government-owned firms. Trends in the last two decades oscillated from general openness toward and even enticement of foreign firms in the 1960s to extremely inward-looking, anti-foreign policies in the 1970s. The trend of the 1980s clearly has been to allow more private-sector participation in the economy, from both domestic and foreign firms. This has not meant much change in ownership of raw materials or utilities, but it has meant increased opportunities for foreign firms to provide contractual services to local state-owned firms and to operate in other industries under fewer restrictions.

Despite these changes in the government policy environment facing foreign business, the outlook in Latin America still remains interventionist. From limits on foreign ownership to price controls and multiple foreign exchange rates, government rules play a central role in shaping and reshaping the competitive conditions in the region. Given the fact that most MNEs originate in industrial countries, they will remain 'foreign,' and thus in some ways threatening, to Latin American host governments. This difference in perspective will not disappear, and it will continue to result in disagreements between the two sides in the years ahead.

The rest of this chapter presents evidence and commentary on the size and composition of foreign business in Latin America during the 1980s. The next section describes the value and distribution of foreign firms in activities such as bank lending, direct investment, exporting, and licensing., The section following that focuses on the nationality of those foreign firms that are undertaking the various activities. The next section then introduces the issue of government–business relations, which is the subject of the next chapter. The concluding section uses the information given earlier in the chapter to sketch some of the dominant trends in multinational business in the region at the end of the 1980s.

The size and scope of MNE activity in Latin America

Foreign bank lending

In value, lending by foreign banks is by far the largest segment of international business in Latin America (and throughout the world). The level of this loan activity has been discussed widely during the past several years of debt crisis. Today the value of foreign loans approaches US$400 billion in total, including both public and private sector borrowing from foreign commercial banks, companies, and official lenders. Table 3.1 gives an overview of the distribution of foreign lending to borrowers in the region.

Foreign debt generally tends to correlate with market size; that is, the largest borrowers are the largest countries. The combined debts of Argentina, Brazil, Mexico, and Venezuela — which make up most

Table 3.1 Indebtedness of Latin American borrowers to foreign lenders, 1986 (US$m)

Country	Total[a] due to foreign commercial banks, all borrowers	Of which, maturities up to one year	Total[a] due from public-sector borrowers to all foreign lenders	Estimates[b] of total foreign debts (1986)
Argentina	30,713	9,568	43,869	53,000
Bolivia	615	446	3,657	4,500
Brazil	67,194	21,640	89,529	109,200
Chile	14,273	5,100	15,491	21,600
Colombia	6,121	2,163	12,476	15,000
Costa Rica	1,441	404	3,894	4,460
Ecuador	5,181	1,846	9,164	9,000
El Salvador	208	108	1,631	n.a.
Guatemala	495	248	2,489	2,640
Honduras	452	294	2,571	n.a.
Mexico	71,508	18,125	80,929	100,400
Nicaragua	499	185	5,615	n.a.
Panama	20,404	13,828	4,710	n.a.
Peru	4,555	2,392	12,346	15,600
Uruguay	2,314	1,233	3,850	5,200
Venezuela	26,859	18,656	26,929	34,100

Sources: Bank for International Settlements, 'Semi-annual International Banking Statistics,' end-June 1986; and World Bank, *World Debt Tables*, 1986–7.
Notes:
[a] World Bank data, estimated at year end 1985.
[b] author's estimates at end-June 1987.

of Latin America's market — constitute just under 90 percent of the total due from Central and South American countries to their foreign creditors. The only major exception to this rule is Panama, a major offshore banking center, which has attracted over $20 billion of 'indebtedness' in the form of eurodollar deposits to its international banking offices. (The foreign debt of Panama, thus, is not reflective of the burden on that country's economy, since most of it is held as dollar balances of foreign commercial banks that are 'parking' funds in the country because of its tax haven status. The World Bank's measure of Panama's official debt, at $4.71 billion, is more indicative of that country's debt problem.)

Although it cannot be seen from these data, a dramatic shift has occurred in the source of foreign lending to Latin America since the 1950s. In a complete reversal relative to previous decades, bank lending is about twice as large as official lending to the region. This situation is a result of the oil crises of the 1970s, when international commercial banks encountered two cycles of excess supply of funds (i.e. new deposits from OPEC members), which they loaned more widely to clients such as Latin American governments, banks, and companies. This moved commercial banks far ahead of foreign government agencies and international institutions as creditors in Latin America. By the end of the 1980s, it appears that once again a larger share of lending to Latin American *governments* (but not private-sector banks and companies) is coming from official institutions such as foreign governments, the World Bank, and the International Monetary Fund. The reversal of lending sources has not been undone, but the tendency since 1982 is for official sources to play an increasing role in financing government projects in Latin America.

The main financial institutions involved in the extension of credit to Latin American borrowers are multinational banks. For example, the amount of cross-border loans owed to Citibank, Bank of America, and

Table 3.2 The major US lenders to Latin America, end–1983 (US$)

Bank	Mexico	Brazil	Argentina	Venezuela	Total
Citicorp	2,900	4,700	1,090	1,500	10,190
BankAmerica	2,741	2,484	300	1,614	7,139
Manufacturers Hanover	1,915	2,130	1,321	1,084	6,450
Chase Mahattan	1,553	2,560	775	1,226	6,114
Morgan Guaranty	1,174	1,785	74	464	4,164
Chemical	1,414	1,276	370	776	3,836
Bankers Trust	1,286	743	230	436	2,695
Continental Illinois	699	476	383	436	1,994
Wells Fargo	655	568	100	279	1,602
First Chicago	870	689	n.a.	n.a.	n.a.

Source: Businessweek, 18 June 1984, p. 21.

Table 3.3 Number of Latin American countries with offices of selected major multinational banks,[a] 1986

Bank (Country)[b]	Branches	With only Rep. offices	With only subs. & affiliates	Total # of L.A. countries covered
Citibank (US) (1)	22	1	1	24
Bank of Nova Scotia (Canada)[c]	12	5	4	21
Chase Manhattan (US)	10	6	2	18
Barclays Bank (UK)	0	0	16	16
BankAmerica (US)	9	4	0	13
Drescher Bank (W. Germany)	1	12	0	13
Algemene Bank Nederland (Netherlands)	0	12	0	12
Bank of Boston Corp. (US)	8	2	1	11
Banco do Brazil (Brazil)	8	3	0	11
Banco Central (Spain)	5	4	1	10
Bank of Tokyo (Japan)	5	4	0	9
Lloyds Bank (UK)	8	0	0	8
Credit Suisse (Switzerland)	3	5	0	8
Bankers Trust New York (US)	2	6	0	8
Swiss Bank Corp. (Switzerland)	0	8	0	8
Compagnie Financiere de Paribas (France)	0	8	0	8
J.P. Morgan (US)	2	3	2	7
Manufacturers Hanover (US)	1	6	0	7
Chemical New York (US)	1	6	0	7
Deutsche Bank (W. Germany)	3	3	0	6
Sumitomo Bank (Japan)	2	3	1	6
Dai-Ichi Kangyo Bank (Japan)	2	4	0	6
Societe Generale (France)	1	3	2	6
Union Bank of Switzerland (Switzerland)	1	5	0	6
Royal Bank of Canada (Canada)	0	0	6	6
Security Pacific (US)	0	6	0	6
Wells Fargo (US)	4	1	0	5
Mitsui Bank (Japan)	1	3	1	5
First Chicago (US)	1	4	0	5
Credit Lyonnais (France)	0	5	0	5
Mitsubishi Bank (Japan)	0	2	3	5
Mitsubishi Trust & Banking (Japan)	0	2	3	5
Sanwa Bank (Japan)	1	3	0	4
First Interstate (US)	1	3	0	4
Fuji Bank (Japan)	1	3	0	4
Industrial Bank of Japan (Japan)	0	3	1	4
Banca Nazionale del Lavoro (Italy)	0	3	1	4
Export Import Bank of Japan (Japan)	0	4	0	4
Commerzbank (W. Germany)	0	4	0	4
Bayerische Vereinsbank (W.Germany)	1	2	0	3
Bank of Montreal (Canada)	1	2	0	3
Daiwa Bank (Japan)	0	3	0	3

Sources: Polk's Bank Directory, 1987; The Rand McNally Bankers Directory, 1984.
Notes:
[a]100 largest commercial banks in the world.
[b]In three or more Latin American countries
[c]All in the Caribbean.

several other money center US banks is shown in Table 3.2. These banks alone have total exposure to borrowers in just four Latin American countries of about $US 50 billion; the same banks collectively have several billion dollars of additional loans exposed in the rest of the region.

This type of lending also is done by hundreds of domestic US, Canadian, and European banks, which do not operate overseas branches or other affiliates. However, the main lenders also operate networks of branches, representative offices, and other affiliates within Latin America, doing local business in markets there. The extent of these networks is shown in Table 3.3, which lists the foreign commercial banks that operate the largest number of affiliates in the region. The largest networks are used by the largest US and Canadian banks, which are located in the industrial countries closest to the region, and the British banks, which historically operated in Latin America to serve their British clients there (particularly in the late nineteenth century). Note that, as of the late 1980s, commercial banks from the other leading industrial countries, Germany and Japan, have begun to establish a substantial presence in the region, though they still trail most of the US money center banks.

Exports and Imports

The evidence relating to MNE participation in Latin American exports and imports is fairly sketchy. While the US Department of Commerce does survey such transactions of US multinational firms, data are not available for firms from other home countries. Recent data show that MNEs account for 62 percent of all US exports to Latin America. Import data are more difficult to interpret, since there are both imports by US MNEs, and also imports shipped by foreign affiliates of US MNEs to the United States. The first category accounted for about 17 per cent of all US imports from the region in 1982, while no data were available for the second category. Apparently, international trade is somewhat more broadly based than direct investment or foreign commercial bank lending in Latin America. Also this is less concentrated than US exports worldwide, where 77 percent of total US exports are carried out by MNEs.

Another aspect of multinational firms' trade is the extent of intra-firm exports and imports. Again data are not widely available on this issue, though the US Commerce Department does track US firms' business of this type. It has been estimated that only 26 percent of US MNE exports to Latin America are to affiliated companies, and only 18 percent of imports to the United States are from affiliated firms in Latin America.[2] From studies done by this author and others, the primary reasons for this relatively low level of intra-firm trade compared to other regions of the world are: (1) government restrictions, (2) transportation costs, and (3) problems of quality control.

Table 3.4 Value of international trade in Latin America, 1986 (US$m)

Exporter	Arg	Bra	Chi	Col	Cri	Ecu	Mex	Per	Ven	USA	World
				IMPORTER							
Argentina	—	575	127	116	3	12	116	120	26	3552	7477
Brazil	797	—	258	134	28	147	197	133	305	15408	22393
Chile	161	293	—	41	na	28	10	66	41	915	4222
Colombia	39	6	35	—	9	42	5	58	80	1854	5102
Costa Rica	7	0	na	2	—	2	1	3	1	541	1125
Ecuador	17	7	54	33	21	—	5	5	2	1457	2183
Guatemala	0	0	0	1	44	7	12	0	1	588	1471
Mexico	99	161	26	103	47	51	—	26	44	11163	16237
Peru	57	74	49	66	5	29	3	—	46	754	2509
Uruguay	71	202	6	7	0	1	8	9	3	442	1088
Venezuela	8	69	116	105	51	5	6	31	—	3764	8412
United States	943	3885	824	1319	483	601	12392	693	3141	—	217307
Total world ($bns)	5.3	15.5	3.1	3.7	1.1	1.9	16.8	2.3	6.9	354.7	1993

Source: International Monetary Fund, *Direction of Trade Statistics*, 1987 Yearbook.

Importer country codes:

Arg	Argentina	Col	Colombia	Mex	Mexico
Bra	Brazil	Cri	Costa Rica	Per	Peru
Chi	Chile	Ecu	Ecuador	Ven	Venezuela

These data can be compared to the size of total international trade in the region, as shown in Table 3.4. While multinational firms apparently do not dominate Latin American trade, nonetheless trade with the United States (and thus with US-based firms) generally outweighs trade with any other nation.

Foreign direct investment

Data on foreign direct investment in Latin America are available from US, UK, and Japanese official agencies in several forms of disaggregation and relating to several aspects of FDI, as well as from the International Monetary Fund for FDI flows on an aggregate basis. The size of FDI flows to the largest Latin American recipient countries is shown in Table 3.5. While Mexico and Brazil are not unexpected as leading recipients of FDI, up until 1985 Colombia and Trinidad are conspicuous as the second- and third-largest host countries and Argentina as only the fifth-largest recipient despite its size as the third-largest Latin American economy. Colombia's position is explained largely by the EXXON coal project that is expected to result in over US$4 billion in total investment during the 1980s. Trinidad's position is the result of that country's oil-producing and refining industry, which involves huge investment by the Royal Dutch-Shell company and other oil-related firms. Argentina's relative unattractiveness probably is due to the economy's great instability

55

Table 3.5 Inflows of FDI to major recipient countries in Latin America, 1970–85 (US$m)

country	Annual averages 1970-4	1975-9	1980-4	1981	1982	1983	1984	1985
Argentina	10.2	119.6	436.3	823.1	225.2	183.9	268.6	967.6
Brazil	851.9	1,820.3	2,103.2	2,525.8	2,922.3	1,556.5	1,598.0	1,281.4
Chile	−141.6	99.1	252.7	383.2	400.8	188.1	77.9	61.9
Colombia	33.9	72.3	367.7	265.3	366.5	617.9	431.5	1,016.4
Mexico	413.1	791.3	1,443.0	2,541.1	1,643.9	454.3	391.6	501.6
Trinidad & Tobago	91.7	106.3	274.5	258.1	345.6	285.0	299.0	−7.1
Venezuela	−140.4	−63.8	125.9	183.9	257.2	85.5	48.2	105.6

Sources: UNCTC, 'Foreign direct investment in Latin America: recent trends, prospects and policy issues,' *UNCTC Current Studies*, series A, no. 3, August 1986, p. 3; International Monetary Fund, *Balance of Payments Yearbook*, 1986.

since 1970 and most recently to the hyper-inflation of 1983–5. The situation shifted somewhat in 1985, when Trinidad's main oil refinery was closed and Argentina rebounded from its debt-crisis decline in FDI.

Another measure that illuminates both the host and home countries involved in FDI in Latin America is the number of affiliates of MNEs from selected home countries in each Latin American host country. Such a measure is presented in Table 3.6. These data demonstrate that US-based MNEs, despite increasing competition worldwide from firms based in other home countries, still maintain dominant positions among foreign investors throughout Latin America. US firms make up over 50 percent of total foreign firms in fourteen of eighteen countries surveyed and over 75 percent of total foreign firms in seven of them.

Licensing, franchising, and management contracting

As noted in Chapter 2, non-equity forms of foreign involvement by MNEs have expanded rapidly in recent years. From simple licensing of technology to complex turnkey ventures, the range of contractual agreements between MNEs and local firms (and other MNEs) is growing at a rapid pace. Some of the contractual agreements supplement existing direct investment projects (such as licensing an affiliate to use the parent's proprietary technology). Other agreements essentially replace an MNE's own production (such as licensing another firm to manufacture a product or franchising another firm to offer a service). Because of the diverse activities included in this category, and the lack of efforts to date to collect extensive information about them, little evidence can be presented at the aggregate level.

The US Commerce Department has been collecting data on US firms' receipts of fees and royalties from foreign sources for a number of years.

Table 3.6 Proportion of affiliates of MNEs from selected home countries in selected Latin American countries, 1980(%)

HOME COUNTRY	Arg	Bhs	Brb	Bol	Bra	Chl	Col	Cri	Dom	Ecu	Slv	Gtm	Guy	Hti	Hnd	Jam	Mex	Nic
Australia	0.1	1.5	1.1	—	0.2	—	0.1	—	—	0.4	—	—	—	—	—	—	0.1	—
Austria	1.1	—	—	—	0.2	—	0.3	0.5	—	0.4	—	—	—	—	—	—	—	—
Belgium	1.7	1.5	—	—	0.9	2.8	0.8	—	—	—	—	0.4	—	—	2.6	—	0.4	—
Canada	1.8	12.0	14.7	1.4	3.0	0.8	0.8	3.0	2.8	1.3	3.4	5.3	4.0	—	6.6	8.9	2.6	3.8
Denmark	0.3	—	—	4.4	0.7	—	0.1	—	0.9	—	—	0.4	—	—	—	—	0.2	—
Germany, Fed. Rep.	9.2	0.5	—	4.4	11.9	9.6	3.7	1.0	2.8	5.7	6.0	3.3	—	—	1.3	1.4	4.2	0.9
Finland	—	—	—	—	0.2	—	—	—	—	—	—	—	—	—	—	—	—	—
France	6.3	1.5	—	—	4.9	6.4	3.1	1.0	—	0.4	0.8	0.4	4.0	2.0	—	0.3	1.8	0.9
Italy	4.8	2.8	—	2.9	2.4	1.6	2.1	—	—	2.2	—	0.4	—	—	1.3	—	0.8	0.9
Japan	1.2	0.7	—	2.9	8.4	1.2	1.0	5.6	0.9	1.3	7.7	2.4	—	—	1.3	—	1.3	1.9
Netherlands	2.2	0.3	1.1	2.9	2.5	2.0	2.3	2.0	6.6	3.0	8.6	3.8	—	6.2	6.6	1.0	0.7	4.8
New Zealand	—	—	—	—	—	—	—	—	—	—	—	—	—	—	—	—	—	—
Norway	—	—	—	—	0.4	—	0.1	—	—	—	—	—	—	—	—	—	—	—
Portugal	—	—	—	—	0.1	—	—	—	—	—	—	—	—	—	—	—	—	—
Spain	1.0	—	—	—	0.3	0.4	0.7	0.5	1.9	0.4	0.8	0.4	4.0	2.0	—	—	0.5	—
Sweden	3.7	—	—	4.4	2.8	4.0	3.5	6.6	1.9	1.7	—	0.4	—	—	—	0.3	1.4	—
Switzerland	3.6	2.8	48.8	1.4	2.4	3.2	1.9	2.0	1.9	2.2	1.7	1.9	—	—	—	0.7	1.8	—
United Kingdom	11.0	28.2	34.0	5.9	9.4	16.8	5.6	—	4.7	8.3	4.3	1.9	64.0	—	1.3	35.6	3.7	4.8
United States	51.2	46.4	—	68.6	47.8	51.2	73.0	77.5	75.2	72.2	66.3	78.2	24.0	89.5	78.6	51.4	79.7	81.5
Other home countries	0.1	0.9	—	—	0.3	—	—	—	—	—	—	—	—	—	—	—	—	—
Total	100.0	100.0	100.0	100.0	100.0	100.0	100.0	100.0	100.0	100.0	100.0	100.0	100.0	100.0	100.0	100.0	100.0	100.0
No. of cases	872	523	88	67	2889	250	563	196	105	227	116	207	25	48	75	278	2349	103

Source: UNCTC Third Survey, p. 327.

Host country codes:

Arg Argentina	Bhs Bahamas	Brb Barbados
Bol Bolivia	Bra Brazil	Chl Chile
Col Colombia	Cri Costa Rica	Dom Dominican Republic
Ecu Ecuador	Slv El Salvador	Gtm Guatemala
Guy Guyana	Hti Haiti	Hnd Honduras
Jam Jamaica	Mex Mexico	Nic Nicaragua

These data give some idea of the size of total contractual ventures by US MNEs, though they ignore such activities of US firms that do not have overseas affiliates. Table 3.7 shows the size of this business in Latin America and world-wide. Note that in all areas of the world, receipts of payments for technology transfers are much greater from affiliated companies than from unaffiliated ones. Unfortunately, it is not possible to separate out the fees received for technology licenses versus management contracts versus franchises, and so on.

Other contractual ventures abound in Latin America as elsewhere, though no summary data are available. Examples of licensing agreements, turnkey ventures, management contracts, co-production agreements, and other contracts can be found regularly in the business press related to the region. (See Business International Corporation's biweekly *Business Latin America*.)

The EXXON joint venture with the Colombian government to mine coal at El Cerrejon in northern Colombia is a good example of complex contracting in recent years. The two parties negotiated a thirty-year turnkey project in which EXXON and the government each own 50 percent of the project during that time. EXXON receives its income from the project in coal for sales outside Colombia, while supplying mining technology, management skills, and some other inputs to the project. Although this description greatly oversimplifies the arrangement, it suffices to show that EXXON has a turnkey project, along with a management contract and several licensing agreements, as well as a countertrade agreement to receive its payments.[3]

The nationality of foreign involvement in Latin America today

In the late 1980s by far the dominant presence of foreign business in Latin America comes from the United States. As seen above, over half of foreign direct investment in the region originates in the United States, as does the majority of foreign bank lending. Aggregate export and import data present a similar picture, though a substantial portion of total transactions are shipped by non-MNEs. That is, US firms still dominate industrial-country trade with Latin American countries, but many of them are domestic US firms with no foreign presence except for the shipment of their products.

This situation is not unexpected, since most multinational business originates in industrial countries, and the one geographically closest to Latin America is the United States. Also, US firms since the turn of the century have been most active overseas through foreign direct investment, banking activities, and other business, relative to firms from other home countries. Excellent transportation and communications links between the Americas continue to favor US (and Canadian) firms over those of other nationalities doing business in Latin America.

Table 3.7 United States: receipts of fees and royalties, 1970–81, 1986 (US$m)

	1970	1971	1972	1973	1974	1975	1976	1977	1978	1979	1980	1981	1986
Receipts from affiliates													
Developed market economies	1,404	1,594	1,815	1,949	2,388	2,770	2,793	3,046	4,054	4,181	4,841	4,805	4,477
Western Europe	811	937	1,088	1,179	1,428	1,765	1,702	1,861	2,561	2,646	3,176	3,035	2,945
Canada	357	389	420	416	541	566	631	673	811	886	931	980	519
Japan	91	103	121	170	211	223	260	302	436	397	413	413	730
Other	145	165	186	183	209	216	200	210	247	253	322	377	283
Eastern Europe	—	—	—	—	—	—	—	—	—	—	—	—	—
Developing countries	491	537	572	519	630	722	686	704	876	1,008	1,227	1,331	344
Latin America and other western hemisphere	318	335	325	269	341	376	299	337	372	422	581	669	183
Asia and Africa	173	202	247	250	290	345	389	367	504	587	646	661	161
Unallocated	24	29	28	46	51	51	51	133	−225	−210	−288	−268	—
Subtotal	1,919	2,161	2,415	2,513	3,070	3,543	3,530	3,883	4,705	4,981	5,781	5,867	4,821
Receipts from unaffiliated foreign entities													
Developed market economies	510	553	583	633	646	641	684	740	859	890	950	1,104	1,542
Western Europe	251	270	274	297	321	344	350	382	423	456	467	590	737
Canada	34	33	38	32	38	38	45	42	47	43	60	64	113
Japan	202	225	243	273	249	219	246	275	344	343	361	379	697
Other	23	25	28	31	38	40	43	41	45	48	62	71	104
Eastern Europe	4	9	8	5	11	14	19	50	34	33	20	22	31
Developing countries	66	64	72	74	94	102	120	130	163	179	215	260	569
Latin America and other western hemisphere	46	46	47	48	63	60	63	71	98	89	101	131	149
Asia and Africa	20	18	25	26	31	42	57	59	65	90	114	129	316
Subtotal	580	626	663	712	751	757	822	923	1,059	1,100	1,185	1,386	2,147
Grand total	2,499	2,787	3,078	3,225	3,821	4,300	4,352	4,806	5,764	6,083	6,966	7,253	6,968

Source: UNCTC Third Survey, p. 356.

British firms in the region have declined in relative importance over the years, but many (e.g. Royal Dutch-Shell, British Petroleum, Barclays and Lloyds Banks, Imperial Chemicals) remain very active in many Latin American countries. In fact, the British share of foreign business in this region relative to that of firms from other European countries is greater than Britain's share in other industrial nations. This may be due to the region's lesser attractiveness relative to European countries and the United States.

The continued British presence follows the nineteenth-century entry of British firms that participated in the construction of railroad, telegraph and telephone lines, and the development of mines, throughout Latin America. Even though these countries are not members of the British Commonwealth (except for some of the Caribbean islands, Guyana, and Belize), British firms have been active in the region for over a century. The Industrial Revolution enabled British firms to profit from their new industrial knowledge in transportation and communications in Latin America just as in the rest of the world.

West German manufacturing firms such as Volkswagenwerk and Mercedes Benz in autos, and Bayer and BASF in chemicals, have long operated in the region with plants as well as marketing offices. The German investment tends to be concentrated, though not exclusively, in the large Southern Cone countries, namely Argentina, Brazil, and Chile. Many other German firms have marketing and some manufacturing presence in Latin America, making Germany the second largest source of foreign involvement in manufacturing in the region.

Canadian firms have established an important presence in the region over the years, principally in raw materials ventures and services such as banking. Alcan aluminum company, Brascan metals company, and a few others still operate some extractive business in the region, often concentrating in the Caribbean. Indeed, two Canadian banks, Bank of Nova Scotia and Royal Bank of Canada, have more affiliates in the Caribbean than any US bank other than Citibank.

Japanese firms only very recently have begun to explore the markets of Latin America. Due to their geographic proximity to the newly-industrializing countries of Asia and other less-developed countries there, Japanese firms did not often look to Latin American countries for new markets or for new production locations. The list of Japanese industrial firms that are active in the region is similar to the general list of Japanese MNEs. In autos, Toyota and Nissan have assembly plants and marketing offices; in computers NEC has offices, and/or distributors in most countries; in consumer electronics, Sony, Panasonic, Mitsubishi, and others sell widely in Latin America. The key differences between these firms and their American counterparts to date is that the American firms have much more experience and many more production facilities in Latin America — but the gap is narrowing.

Swiss firms such as Nestlé. Hoffmann-La Roche, and Ciba-Geigy, compete actively in Latin America. Since none of these firms is in a heavy, capital-intensive industry, their size is much less than that of MNEs from other countries in extractive or automotive businesses. None the less, the major Swiss MNEs do operate through some local assembly plants, some marketing affiliates, and typically many distributors.

In all, the distribution of foreign companies active in Latin America follows generally the pattern of these MNEs in the industrial countries, with relatively greater participation from the geographically closer US and Canadian firms and relatively less by the distant Japanese firms.

The relevance of goverment policy to distribution of activities

Exclusion of foreign firms from some sectors

An overview of total foreign business ventures in Latin American countries shows the concentration in the largest countries, especially Brazil and Mexico, ownership predominantly by US firms, and sectoral distribution that favors large raw materials ventures, manufacturing facilities, and services such as banks, hotels, and restaurants. This distribution of activities is determined primarily by the size of the host country market, but also importantly by the regulatory environment. Since the early 1970s' nationalizations of oil industries in many countries in all parts of the world, that sector has declined in relative importance to foreign business. None the less, petroleum remains the single largest sector for foreign business in Latin America today.

Similarly, nationalizations in copper and bauxite have expanded local company roles in those industries in Latin America, but some foreign companies still operate in all of these sectors, and their presence often is very large. With this small number of highly visible exceptions, the rule clearly is that foreign business is most welcome and most active in manufacturing and service industries during the 1980s.

By far the number of foreign firms is greatest in services, when those services are defined to include the sale of manufactured goods imported from abroad. Far more manufacturing MNEs operate sales offices in Latin American countries than manufacturing facilities. Other services such as commercial banking often are limited to local institutions, as are insurance and public accounting. The ownership of public utilities such as electric power provision, telephone company operation, and postal service is held by the government in every Latin American nation. In these industries, foreign MNEs participate as vendors of equipment and management skills to the government firms in each country. Thus for the foreign firms these also would be classified as service activities.

State-owned firms

Even when foreign ownership in an industry is not precluded by sectoral bans, as in petroleum production or telephone service, the existence of state-owned companies tends to deter foreign business. For a fairly common example the petrochemicals industry often has extensive participation by the state-owned oil company. Usually this firm does not dominate the chemicals industry, but it restricts some areas to itself and is able to acquire government protection for those segments that it wishes to serve. (This phenomenon is discussed in more detail in Ch. 11 below which looks specifically at the chemicals industry.)

Table 3.8 Percentage of government participation in the economy, 1985[a]

Country	0	10	20	30	40	50	60	70	80	90
			% of GDP							
Argentina			18%							
Bolivia					40%					
Brazil			21%							
Chile					36%					
Colombia	n.a.									
Costa Rica			21%							
Ecuador		15%								
El Salvador			20%							
Guatemala		13%[b]								
Honduras	n.a.									
Mexico				25%						
Nicaragua						49%[b]				
Peru		13%								
Uruguay				25%						
Venezuela				26%						

Source: World Bank, *World Development Report*, 1987, Table 23.
Notes:
[a] These measures exclude the output of government-owned companies, except those that provide government services such as defense, education, health, housing, social security, economic services, and general administration.
[b] Percentages for 1983.

Government ownership of industry is fairly large throughout Latin America. All of the countries have over 25 percent of GDP generated by the government sector (including government-owned firms).[4] Table 3.8 shows the approximate percentages of government participation in the Latin American economies. One of the key implications of this extensive government participation in business is that private corporations, both domestic and foreign, must be aware that their competitors and/or suppliers and/or customers are government-controlled, and thus potentially able to obtain government protection to pursue their own interests. Viewed in another way, over one-fourth of the business done in Latin American countries involves a government, which often can pursue non-economic goals and employ non-competitive strategies that place the private firm at a disadvantage. This issue of the competitive environment and government–business relations is pursued in more detail in the next chapter.

Markets

Having accounted for the two major factors that determine market structure other than competition — namely precluded industry sectors and state-owned companies — the discussion can focus on the main determinant in most instances, namely competition.

Competition in Latin American countries tends to be limited by both government involvement and small market sizes. Except for Brazil and Mexico, the countries in this region can be considered small and medium-sized LDCs, with underdeveloped markets characterized by small pockets of industrialization and large expanses of rural subsistence-level farming. This type of market generally leads to optimum-scale output for only a small number of firms in many industries. Thus competition tends to be oligopolistic and government regulation is important to prevent abuse of dominant (monopolistic) positions in those industries where scale economies are important.

The relative irrelevance of incentives

One of the most interesting and important findings of studies on the use of incentives by LDC governments in recent years is that enticements such as tax holidays or subsidized loans have negligible impact on company direct investment decisions.[5] Such decisions depend primarily on market size and competitive conditions in the host country. While restrictive government policies such as limits on foreign ownership tend to depress direct investment marginally, other policies appear to have virtually no impact at all. Repeatedly, managers of foreign firms state that preferential policy is not so relevant a consideration as *stability* of government policy. That is, firms want to avoid uncertainty about government policy changes rather than receive special treatment (which may be taken away by a subsequent government).

What's next?

International economic conditions in the late 1980s portend a fairly bleak near-term future for Latin America. Commodity prices appear to be stable at historically low real values. Industrial country markets are becoming more difficult to penetrate, not only because of greater competition but also because of somewhat increased protectionism. The problems of operating largely state-controlled economies are not decreasing. This situation consequently should make the government policy environment more favorable to the MNEs seeking to operate in Latin America. On the other hand, a weak local economy will tend not to attract foreign investors in the first place, if other, more dynamic economies are available as alternatives.

For the large MNEs that have substantial experience in Latin American countries, the late 1980s may well be a time for increased investment (financed as much as possible through local retained earnings and borrowings). With the expectation that conditions in the region will improve as they have in the rest of the world since the severe recession of 1980–3, MNEs should see markets such as Brazil, Mexico, and Venezuela as desirable targets for future business and thus current investment.

For non-US MNEs, with a substantially-devalued dollar in the late 1980s, investment in the 'dollar area' is relatively inexpensive. Given that German and Japanese firms already have established major competitive positions in the European Community and the United States, they are very likely to build their bases in areas such as Latin America. As usual, the larger markets will attract the greater investment and other business activities; so Argentina, Brazil, Mexico, and Venezuela can be expected to receive growing attention from the German and Japanese firms.

One politically important but economically weak part of the region-that has been largely ignored here is Central America. If by some means the political instability problems can be solved from Guatemala to Panama, these countries stand to gain important inflows of foreign business, from small-scale manufacturing operations to serve local markets all the way to offshore-assembly plants to serve the North American market. Although this outcome does not appear imminent, it is geographically sensible, and it awaits some resolution of the chronic instability of the Central American countries.

Finally, it should be noted that the foreign debt crisis has still not (in 1988) been averted in Latin America. The specter of multibillion dollar debt service payments hangs over the whole region so importantly that it has effectively discouraged growth in long-term investment for most of the decade. This crisis may be in the process of resolution with the development of a secondary market in foreign loans, use of debt-equity swaps, and other mechanisms for eliminating loans from both bank and government books — and finally accepting the losses involved. The results still remain to be seen, and the attitudes of MNE managers still remain dubious of major recovery in the next few years.

Notes

1. United Nations Center on Transnational Corporations, 'Foreign direct investment in Latin America: recent trends, prospects, and policy issues, *UNCTC Current Studies*, Series A, no. 3 (August 1986), p. 1.
2. These data come from the *Survey of Current Business*, various issues, and from the US Department of Commerce, *US Direct Investment Abroad: 1982 Benchmark Survey Data*. Washington DC: USGPO, 1982.
3. For more details, see EXXON Corporation, *The Cerrejon Project*, INTERCOR: Bogota, Colombia, 1982.
4. The percentage of GDP generated by the government sector is actually substantially larger in the European Community countries than in Latin America. The average in Europe is just about 50 percent.
5. See, for example, Steven Guisinger, *Investment Incentives and Performance*, New York: Praeger, 1985; Robert Grosse, *Foreign Investment Codes and the Location of Direct Investment*, New York, Praeger, 1980, Ch. 3.

Bibliography

Guisinger, Steven (1985) *Investment Incentives and Performance*, New York: Praeger.
United Nations Center on Transnational Corporations (1983) *Transnational Corporations in World Development: Third Survey*, New York.
United Nations Center on Transnational Corporations (1985) *Recent Trends in Direct Foreign Investment*, New York.
United Nations Center on Transnational Corporations (1986) 'Foreign direct investment in Latin America: recent trends, prospects, and policy issues,' *UNCTC Current Series*, series A, no. 3 (August).

Part two

The Regulatory and Economic Environments

Chapter four

The government-business relationship in Latin America

Introduction

Latin American statism

To multinational firms operating in this region, government-business relations play a far more important role than in any of the industrial countries where MNEs typically produce and market their products and services. It is not just that the government (and government-owned company) sector comprises more than one-fifth of the total economy throughout Latin America. This implies that the government in each country is at once a major competitor, customer, and supplier, as well as being the rule-maker. This last point is central. It cannot be overemphasized that dealings with the government, all the way from obtaining necessary permissions and approvals to paying taxes and complying with other statutory provisions, are crucial aspects of competition in Latin America.

The size of the government sector, including the bureaucracy itself plus government-owned firms, the military, and so on, is fairly similar to other non-communist countries of the world. Table 3.8 has already shown estimates of the proportion of total GDP generated by government-related activity (except state-owned companies) in selected Latin American countries in 1985. Ranging from a high of Nicaragua's 49 per cent (ignoring Cuba) down to a low of Guatemala's 13 percent of GDP (again, excluding state-owned companies), the level of participation of the government in every country's economy is important. To put this in perspective, the Latin American situation contrasts with the degree of government participation in the economies of Western European countries, where the average government sector is closer to 50 percent! Even adding in the state-owned companies, we find that they account for typically 10-20 percent of GDP, still leaving Latin American governments as less directly involved in generation of national income than the governments of Western Europe. Thus, looking simply at the size of government[1] does not illuminate the full nature of the role played by governments with respect to business in the region. Another view is needed.

First, the economic 'model' on which most Latin American countries operate is some type of combination of capitalist and socialist, with substantial room for private-sector business but also a fundamental commitment to allowing the government to limit business activity and direct the economy. This kind of 'statism' has lead over the years to an interdependence between the private sector and the government which has proven quite resistant to change. Howard Wiarda reasons that, in Latin America, 'The state is viewed as the prime regulator, coordinator, and pacesetter of the entire national system, the apex of the Latin American pyramid from which patronage, wealth, power, and programs flow.'[2] In this context, successful dealings with the national government are a prerequisite to business success in the country.

A fascinating example of this government–business relationship involving the Mexican government and the major US auto firms is offered in Bennett and Sharpe (1985). They trace the process of entry by major US auto companies, and focus on the firms' negotiations with the Mexican government between 1960 and 1980. Their 'historical-structural' approach emphasizes the bargaining relationship between companies and successive Mexican governments, showing how the strengths and weaknesses of each side led to outcomes in negotiations over time.

Second, the underlying reason for the statist tradition may be read from Latin American history, which shows the consistent reliance on governments to choose and enforce appropriate economic direction for the national economy (cf. Wiarda and Kline, 1985). This direction, in turn, was a traditional characteristic of the Western European colonial powers, Spain and Portugal, that settled Latin America. While the Latin American countries for the most part have not placed as much of their economies directly into government agencies, these countries have kept control over economic activity through other policy avenues such as licensing regulations, other permissions, and a wealth of specific statutes that guide business as the governments choose — as well as extensive use of government-owned companies, especially in the huge, capital-intensive industries such as petroleum, copper, bauxite, and other minerals and metals.

Thus the size of the government sector ought to be measured in terms of the government's influence on the daily operation of business in Latin America, to give a realistic picture of the importance of governments in determining business activity in the region. This influence includes the direct participation of government agencies and companies in the economy, plus the legal and policy framework that a government places on activities of other businesses. The bulk of this chapter is devoted to exploring the ways in which Latin American governments try to channel and control business activities of foreign MNEs.

A stylized sketch of the emergence of a new firm may convey more concretely the idea of governments' channeling (foreign) business into desired activities. Assume that a new company is formed and begins to compete in one of the sectors not reserved to the government in a Latin American country. Once the firm establishes its niche in the local market, and creates its contacts with government bureaucrats who provide the necessary permissions and other paper work necessary to function under the law, that firm tends to become a part of the system. The system, in turn, can make it difficult for subsequent firms to enter the same market niche as the established one. The initial firm then becomes more committed to the government. This self-reinforcing system of government–business relations obviously is different from what MNEs face in many countries. In fact, very often it is difficult for MNE managers assigned to Latin American posts to comprehend this 'system' except through their own personal experiences over time.

Thus the relationship between a Latin American government and a foreign MNE seeking to establish operations in that country depends largely on the existing ties between that government and local firms or other MNEs with existing facilities there that would compete with the new MNE. Unless the MNE offers some clearly superior benefits to the government in comparison with the established local firm(s), it is likely to be quite difficult for the new MNE to enter that market. If the MNE seeks to enter a market segment not served by local firms, then the probability of successful entry would be much greater, *ceteris paribus*. Obviously, this latter strategy is desirable, but most times the market is sufficiently developed to have existing local competitors, in which case very careful attention must be paid to establishing good (i.e. mutually beneficial) relations with the host government.

A history of government–business ties

The environment described above is not a phenomenon of the recent past. Since the independence of most of Latin America in the early 1800s, governments have tightly controlled their economies. As far back as the 1930s the Mexican government chose to restrict activity in the petroleum sector to a state-owned firm (Pemex) and to nationalize multinationals in that sector. Similarly, public transportation (especially railroads) and telecommunications have been nationalized by Latin American governments since the turn of the century.

Another aspect of the historical ties of governments to business in the region are links between family-owned business conglomerates and governments. In each Latin American country there are small numbers of families whose business activities have expanded to encompass large parts of the total economy. For example, in Venezuela, members of the Diego Cisneros family own controlling interest in Cada supermarkets, Pepsi Cola y Hit de Venezuela, Venevision TV station, and several other diversified industrial concerns. Similarly in Peru, the Nicolini family owns the main flour industry, chicken and egg production, and other food industries. They also own insurance companies (such as La Colmena) and are the main shareholders of several other financial institutions.

Each of these family conglomerates extends its ties to the local government through family members who work as government officials, through years of compromise-building on projects and laws which affect family enterprises, and through participation of family members in political parties and other key interest groups. Any new firm entering a business segment already occupied by a family concern is likely to face serious opposition if it threatens to disrupt the status quo.

MNEs with competitive and bargaining advantages

Given these historical underpinnings of the system, foreign multinational firms have often found the business environment in Latin American countries to be hostile, or at least difficult to understand. Due in large part to the continued acceptance of the statist tradition, this environment is an enduring aspect of the region, which shows few signs of changing in the 1980s. What the MNE manager needs, then, is to understand the issues on which governments are likely to be cooperative and those on which they are likely to be conflictive. On this basis, the manager can consider the firm's strengths and weaknesses relative to the government, and then can develop strategies that take best advantage of the firm's strengths.

A view of the bargaining advantages of both MNEs and host governments was presented in Chapter 2. One of the goals of the present chapter is to explore the importance of these advantages in determining the policy environment in Latin America today. At the same time, it should be remembered that success in these markets also depends on the competitiveness of the firm relative to its rival firms. So the issue of competitive advantage will play an important part in company strategy along with bargaining advantage.

The rest of the chapter is laid out as follows. First, the bargaining theory is reviewed, and a number of additional views are examined. Second, some of the most important regulatory issues and policies in specific countries are discussed and compared between countries. Third, an effort is made to test some of the implications of the bargaining theory in terms of regulation of different kinds of firm and industry. Finally, the bargaining theory is reevaluated in light of the results of the tests.

More on the bargaining theory of government–business relations

The bargaining theory has been presented as a kind of game theory, with two central actors (the national government and the MNE) but also including the non-trivial participation of more actors, such as competing firms, local governments, and foreign national governments. The game is one of resource allocation and benefit distribution between the government and the company, though with direct impact on the firm's customers, as well as on people in general in the country at issue. What we would like to do in the present context is to measure the relative importance of various factors in the determination of the game's results — in other words, to explore the factors that will improve the government's net benefits and those that will improve the firm's benefits from a bargaining situation. Before turning to that task, two other, similar perspectives are presented.

In their book *Multinationals under Fire (1980)*, Gladwin and Walter sketched a model of bargaining relations that was adapted in Figure 2.1. The central idea of their analysis is that company–government bargaining results are dependent on several dimensions of the relationship. Most obviously, the results depend on the relative power of company and government in the relationship. For example, if the company makes pocket calculators, the government has a fairly strong bargaining position if its national market is large: there are many actual and/or potential suppliers of pocket calculators, there are not huge scale economies in producing that product, and generally the firm does not have a strong proprietary position. On the other hand, if the product is the latest supercomputer, there are very few suppliers, the technology is to some extent proprietary, scale economies are very large, and in this case the government has much less bargaining power.

In addition to the dependence on relative power, the bargaining relationship has another key dimension: the importance of the activity to each party; If the project is a petrochemical plant that a small country wants to have the MNE construct, this plant may be a major part of the country's economy but a small part of a multinational oil company's business. In this situation, the firm has a stronger bargaining position.

Another key dimension is the interdependence of each party's interests. If the government and company stand to gain significantly from the success of a project, then both will be accommodating toward the other to assure that success. If the project will greatly benefit one party but not benefit the other, then the bargain favors the party that will gain less.

A fourth and final dimension discussed by Gladwin and Walter is the quality of the relationship between company and government. In their study they found that the existing relationship between the government and the company had an important impact on the results of a bargaining situation. If the two sides had dealt previously and had achieved a relationship of mutual trust, then a successful bargain was more likely. If the government had previously had trouble with the firm, for political, social, economic, or other reasons, then a bargain was much less likely to succeed. This characteristic is not really a bargaining dimension that favors one or the other, but a factor that affects the overall likelihood of reaching a successful negotiation. For this reason we ignore the relationship quality in the discussion below.

Another, similar perspective is offered by Theodore Moran[3]. He sees five factors that will tend to determine the outcome of a government–company negotiation, looking specifically at a direct investment project. First, the general economic attractiveness of the project: the more attractive the project, the more accommodating the government will be. Second, the size of the fixed investment: the larger the investment, the more accommodating the government will be and the more demanding the company will be. Third, the rate of change of technology involved in the project: the faster the rate, the better the bargaining position of the company that is creating the technology. Fourth, the role played by marketing: the more marketing-intensive the product, the stronger the position of the firm, since the firm controls the marketing skills and brand or trade names. Fifth, the degree of competition in the sector: the more companies in the sector, the better the bargaining position of the government, which can choose among alternative suppliers.

Moran's view of the bargaining situation is not at all at odds with that of Gladwin and Walter; rather, he adds some new considerations that do in fact affect the expected outcome of government–company negotiations. In particular, his third factor, the rate of change of technology, was quite important in determining the bargaining successes of many of the firms studied in this analysis.

Before measuring the impact of various factors on the bargaining outcomes, let us examine the existing regulatory environment in Latin America, with examples and comparisons of policies in different countries of the region.

Current rules on foreign business in Latin America

This section provides a brief overview of the legal environment facing foreign firms that seek to operate affiliates in Latin American countries. Only a selection of these laws is presented; a more current and complete review is available in publications such as Business International Corporation's country analyses in *Investing, Licensing, and Trading Conditions*, or the Price Waterhouse series of *Guides to Doing Business in* [each country].Of course, even more precise information is available from local sources in each country. The focus here is on those rules that have been found most important to MNE managers in the region during the past ten years.

Laws and regulations affecting foreign direct investment

The broad perspective of host governments in Latin America during the 1970s and 1980s is that foreign MNEs constitute an important link to industrial technology and industrial-country markets, but at the same time these firms must be watched and regulated to ensure that the host country gains the most benefit possible from their activities. Since the late 1960s, a large body of restrictions on foreign companies' ownership, pricing, financial transfers, and other structures and activities has been created. While the policies on each government's books appear to be fairly restrictive, in many cases exceptions are made and the actual operating environment is not nearly as limited as it may appear. Table 4.1 compares selected regulations on the activities of foreign MNEs in nine Latin American nations.

The members of the Andean Pact clearly presented the most formidable legal roadblocks to foreign MNEs seeking to operate there. (Ch. 6 explores the Andean rules in more detail.) Until mid-1987 most, though by no means all, foreign investors were being channeled into joint ventures with local partners, as a means of forcing some of the control and profits of foreign firms into local hands. While Mexico has followed a similar policy since 1973, none of the other Latin American nations demand local ownership.

In addition, the Andean Pact members limited from (1971–87) financial transfers by MNEs, toward the goal of preserving scarce foreign exchange and avoiding even more speculation against their currencies. Given the continuing cloud of foreign indebtedness throughout the region, most other countries also have been forced to constrain access to dollars and other 'hard' currencies during the 1980s. Exchange controls are the rule rather than the exception during the present decade. None the less, the Andean countries still exceeded other countries in the region with their restriction of profit remittances and disallowance of royalty payments to parent firms.

Other areas of regulation that particularly affect MNEs are local content (including local labor) requirements, which are used to attract local production by firms that would otherwise import products from their foreign facilities. Local labor rules in Latin America are not very different from those in the rest of the world, in the sense that foreign nationals need permits to work locally, and such permits are issued typically only to highly skilled foreigners whose specific knowledge is not available locally. Beyond this, however, many countries in the region require a percentage of the sales value of products sold locally to be paid out locally, to local workers and suppliers who contribute to the value added of the product. These percentages range widely, with some even demanding 100 percent local production, for example, in Brazil for microcomputers and in Venezuela for some foods.[4]

Table 4.1 Selected rules affecting affiliates of foreign MNEs

Country	Ownership	Profit remittance	Technology transfer	Exchange controls	Local content	Price controls
Argentina	Unrestricted	Supplementary tax on remittances above 12% of registered capital	Unrestricted; pharmaceuticals not patentable	extensive controls	80% required on autos	Repeated price freezes since 1985; most products controlled
Brazil	de facto govt pressure for majority Brazilian ownership	supplementary tax on remittances above 12% of registered capital	licenses must be registered; royalties allowed of 1–5% of sales	2-tiered exchange market; access to official rates is heavily controlled	on many products; over 90% on autos	repeated price freezes since 1986; most products affected
Mexico	49% foreign is maximum	unrestricted except by general exchange controls	licenses must be registered; royalties allowed up to 7% of sales	2-tiered exchange rate	On many products; 60% on autos	3-tiered system of limits, on most products
Andean Pact Bolivia	Decision 24	Decision 24	Decision 24	2-tiered exchange market	n.a.	On some products
Columbia	"	"	"	Strict	Strict	On 30 categories of products
Ecuador	"	"	"	Fairly strict	Required in autos	On drugs & some foods
Peru	"	"	"	2-tiered exchange rate	25–50% in many sectors	On all products
Venezuela	"	"	"	3-tiered exchange rate	Required in many sectors	On 150 categories of products

Sources: Business International Corporation, *Investing, Licensing, and Trading Conditions*, 1987; Price Waterhouse, *Guide to Doing Business in Bolivia*, 1985, New York: Price Waterhouse.
Note: Decision 24 of the Andean Pact required fade-out of foreign ownership in foreign direct investment projects to a maximum of 49 per cent over a fifteen-year period. Also, profit remittances were limited to 20 percent of registered capital per year, and payments for technology transfer from parent to subsidiary firm are not allowed. These rules are *not* strictly enforced in any of the countries, and many exceptions are allowed. Decision 24 was formally dropped by the group in 1987.

Up to this point, all of the rules discussed have been restrictions. In every country there also exist a variety of incentives to attract firms into government-selected industries and locations. Typically, tax reductions and/or low-interest loans are offered to firms that set up factories or other facilities in locations away from the major population centers. This policy seeks to distribute economic activity more evenly through the country and to alleviate the urban crowding that exists in population centers. Also, particular industry sectors often are chosen for stimulation, and MNEs can take advantage of helpful legislation that supports the local agricultural industry (Venezuela), assembly of manufactured goods (Mexico), or oil exploration (Peru).

Trade policies

Though generally not limited to multinational firms, trade policies in each country play an important part in the regulatory environment, too. In most of the region, imports of some products are severely limited through licensing and prior-deposit requirements. Imports also are constrained simply because of difficulties in gaining access to foreign currency (most importantly, US dollars). While black markets exist

Table 4.2 Import restrictions in Latin America

Country	Tariffs	Non-tariff barriers	Other
Argentina	Range from 0–60% ad valorem	Imports must have certificate of necessity from DJNI. Govt imports must be on Argentine-flat vessels.	n.a.
Brazil	Range from 0–400% ad valorem	Anti-dumping duties. Govt purchases should be from suppliers or imported on Brazilian vessels.	Firms are pressured to maintain positive balance of payments.
Chile	10% uniform tariff ad valorem	Import payment must be made after 120 days after shipment; license required on all imports.	Used cars may not be imported.
Colombia	Maximum 80% ad valorem	License required on 33% of products imported; quotas on raw materials.	Prior deposit required for import license.
Ecuador	Range of 0–200% ad valorem	License required for most imports; 180-day prior deposit required.	
Mexico	Range of 0–45% ad valorem	Imports of luxury cars and vans prohibited. Advance deposit of full duties on imported motor vehicles.	1/8 of imports are subject to quotas, mainly chemicals.
Peru	Average 57% on mfg. 25% on raw materials ad valorem	300 products prohibited; license required on all imports; no prior deposit.	n.a.
Venezuela	Range of 0–500% ad valorem	License required for many products: coffee, salt; some clothing articles prohibited.	n.a.

Sources: Business International Corporation, *Investing, Licensing, and Trading Conditions* New York; Price-Waterhouse, *Guide to Doing Business in Chile*, New York (current editions).

The regulatory and economic environment

everywhere that dollar availability is limited in the official market, the price of those dollars often is exorbitant, sometimes reaching more than 100 percent above the official rate. Table 4.2 highlights some of the import restrictions used in Latin America today.

Generally, restrictions are greatest on products that are also produced locally. To foster development of specific industries (e.g. auto assembly), governments target a few additional products for import limitation as well. The rules do not appear more severe in any one country or sub-region, though Chile stands out as the least regulated import market.

The importance of state-owned firms

Another important aspect of government policy that affects foreign MNEs in Latin America is the operation of government-owned companies. In most of these countries, the largest firms include the national oil company, other natural resource companies, and public utilities. These enterprises typically preclude MNE entry into some activities (e.g. refining oil or providing telephone service), though they usually deal with MNEs as suppliers and as customers. Since state-owned companies tend to be concentrated in natural resource industries and utilities, they tend not to alter the competitive conditions in other industries. Chapter 12 explores the impact of these state-owned firms in more detail.

Testing the bargaining theory

The bargaining theory posits that both government and company are interested in maximizing their own gains from the use of resources in business activities. By focusing on each party's key advantages in the process of ongoing government-business relations, some useful conclusions can be drawn. The analysis is based on bargaining situations

Table 4.3 Eight major bargaining advantages

	Bargaining advantages	Favors the firm or the government	Expected regulation
1.	Proprietary knowledge	Firm	Low
2a.	Marketing skills	Firm	Low
2b.	Export dependence	Firm	Low
3.	Scale/scope economies	Firm	Low
4.	Ease of moving facilities	Firm	Low
5.	Dependence on local resources	Government	High
6.	Dependence on local markets	Government	High
7.	Public visibility	Government	High
8.	Competitiveness of the industry	Government	High

involving direct investment projects, licensing agreements, export-import activity, and other international business as information is available. From a review of the relevant literature, four dimensions have been identified in which the MNE possesses a bargaining advantage and four in which the host government has an advantage. Table 4.3 lists the eight power dimensions that alternately favor company or government, plus an additional dimension the impact of which changes over time. Each dimension is discussed as a hypothesis of expected outcomes of company-government negotiations, and empirical evidence is drawn upon to support the hypotheses.

Company advantages

First, the power of the firm is greater in situations where its *proprietary knowledge* is more important (such as pharmaceuticals, computers, other electronics). By contrast, if the technology involved in the project is fairly mature or standardized (as in electrical appliances, foods, and other agricultural products), then the firm has less bargaining power with the government, since the knowledge readily could be obtained elsewhere.

This hypothesis is supported most directly by the structure of the production process in high-tech industries. Very little basic production of computers, pharmaceuticals, telecommunications equipment, medical equipment, and other products that are based on very recent advances in technology is carried out by MNEs in Latin America. Firms such as IBM, Burroughs, Digital Equipment and other computer makers operate many sales offices in the region, but virtually no production. (IBM does assemble PCs and microcomputers in Mexico and minicomputers and mainframes in Brazil.) Similarly, the pharmaceuticals MNEs use local sales offices to serve Latin American markets, plus some formulation facilities, which function as final assembly plants for pills, lotions, tablets, and so on. Pharmaceuticals research and basic production of chemical entities used in the final products is done outside of Latin America. The same production structure exists in telecommunications and medical equipment, with very little production done in Latin America, except some assembly of products that typically incorporate well-known technology.

Another simple measure of the ability of high-tech firms to avoid constraints imposed by Latin American governments is their ownership structure. One would expect that such firms can achieve full ownership of their affiliates, since the Latin American governments need their products. In a survey of Fortune 500 firms that operate in at least two of the largest four Latin American countries, Grosse (1988) found that percentage ownership of MNE affiliates was positively correlated with a firm's expenditure on R & D — a common indicator of technology intensity. The test statistic, however, was not significant for this sample.

Pearson's $r = 0.15$ for the sample of fifty-five firms in 1986. In another study of US-based MNEs, Fagre and Wells (1982) discovered that firms with very high expenditures relative to sales (greater than 5 percent) did indeed achieve significantly greater ownership of Latin American affiliates than other firms, though they also found that medium-tech firms had no better results than low-tech firms in their sample.

Finally, the number of exceptions to host-country rules on ownership, technology transfer, profit remittance, taxes, and so on, includes a greatly disproportionate number of high-tech firms. In 1986 IBM received permission to construct a new, wholly-owned production facility in Mexico, despite the long-standing rule (since 1973) that requires majority ownership by Mexican investors in any local companies. Gereffi (1985) discovered that the Mexican government was unable to constrain foreign drug companies producing steroid hormones to sell part ownership to Mexicans or to further the goals of the government in increasing the country's exports of these products. The foreign MNEs controlled the technology and the market access abroad (and they had access to a needed raw material from other countries as well).

Second, the power of the firm is greater in situations where *marketing skills* play a larger part in the business. That is, if the product to be produced is a branded good (such as a washing machine, television, or automobile), the firm which possesses the brand has a stronger position than any other firm in the same business. By contrast, if the product to be produced is not a branded good (such as standard base chemicals, foods, and metals), then the company's brand name carries with it little bargaining power.

This hypothesis is most readily supported by examples in capital goods industries. For example, manufacturers of autos, tractors, and other transportation equipment have historically received numerous concessions from host governments to attract investment in assembly plants. Typically, the small Latin American markets (other than Brazil and Mexico) do not justify optimum-scale production facilities, but government incentives are used to attract the foreign MNEs anyway.

Measuring marketing intensity as 'advertising expenditures relative to total sales', another set of tests can be used to determine if marketing skills do correlate with lower government regulation. Comparing firms from the Fortune 500 sample on the basis of advertising expenditures versus percentage ownership of Latin American affiliates yields a Pearson's $r = 0.34$. In this sample, the test statistic is significant at the 99 percent confidence level, implying that marketing intensity *is* an important source of bargaining advantage.

Another aspect of marketing advantage relates to exporting. The bargaining power of the firm is greater when the firm possesses a more extensive network of overseas affiliates. That is, the more it can gain access to foreign markets for exports from the host country in which the bargaining takes place, the better treatment to be expected from the host government. As in most countries today, if the government wants to stimulate exports, then MNEs with large affiliate networks have a bargaining advantage. If the MNE does not have superior access to foreign markets, then the government has greater power to dictate terms to that firm in order to secure foreign sales. The Gereffi study (1985) cited above corroborates this view.

In the period of region-wide debt crisis since 1982, MNEs operating in Latin America have encountered a barrage of government incentives that benefit firms that export a large part of their output. In fact, to obtain reasonably assured access to foreign exchange, many firms have been pushed to arrange for extensive exporting in return for this privilege. Caricom countries, for instance, provide a special incentive regime for exporters, based on a generous income tax and duty reductions. Subsidies and tax rebates apply in Colombia and Peru for non-traditional exports. Brazil and Paraguay offer also incentives for investment in export industries.

Third, the power of the firm is greater in situations where *economies of scale or scope* are important (as in producing automobiles or airplanes, refining oil, or making various steel shapes). In each case, the financial and knowledge resources are quite large, and typically are possessed only by a few firms. At the other end of the spectrum, there are businesses with very low scale economies (such as production of many foods, cosmetics and health care products, and clothing) in which MNEs possess no special size advantage relative to other local or foreign firms. In these latter cases, the host government has a bargaining advantage because of the availability of alternative suppliers (of presumably similar quality).

This hypothesis can be explored by comparing the size of a foreign investment project with the percentage ownership of the parent; the larger the scale of the project, the greater the expected percent ownership. In this case, using the previous sample, the Pearson correlation coefficient is 0.10, which is not significant.

Fourth, the power of the firm is greater in situations where *the firm is relatively 'footloose'*, that is when it can easily shift its facilities from one location to another. This condition exists in many service industries, when the service is legally structured to take advantage of a low-tax location or other unrestrictive environment (e.g. in banking, insurance, and other financial services). It also occurs in the case of offshore assembly in manufacturing. If the host country creates an onerous legal burden, the firm is likely to move its assembly operation elsewhere (e.g. from Mexico to the Far East). By contrast, many industries are not at all footloose, and the obsolescing bargain strongly favors the host government (e.g. oil refining, auto manufacturing, provision of services such as tourism).

This hypothesis is readily supported with evidence from the banking industry. Every time a major change takes place in national laws that relate to international banking, deposits, loans, and/or people are shifted to more desirable locations. When the United States created an onshore eurodollar market (through International Banking Facilities) in 1981, billions of dollars of deposits moved from Panama and Nassau to Miami and New York. When Panama created a tax-free haven for international banking transactions in 1970, billions of dollars and many bank offices moved there in immediate response. In fact, the history of growth and change in the banking industry world-wide is largely shaped by regulatory changes rather than shifts caused by other factors.

Considering other industries that are involved in manufacturing rather than service provision, somewhat similar results can be found. The phenomenon of offshore assembly of textiles and electronics products has boomed in the Asian newly-industrializing countries (Hong Kong, Korea, Singapore, and Taiwan), but also importantly in Mexico, Jamaica, and the Dominican Republic. Particularly in the Mexican case, the growth of this industry occurred directly as a result of a Mexican legal change that permitted the 'maquiladora' zones to operate tax-free. The Jamaican and Dominican cases appear to be more the result of these countries' proximity to the United States, their political stability, and the lack of legal constraints (rather than important incentives). None the less, it is the legal climate, and its stability, that attract offshore assembly to particular countries in a region where labor costs are generally low everywhere.

Government advantages

Fifth, the power of the government is greater in situations where the industry is based on a *raw material* available in the host country (such as oil, copper, or a tropical climate). Since the host government has ultimate sovereignty in any country, the more dependent the MNE on some resource of that country, the more powerful the government's bargaining position. At the other extreme, if the firm uses very little or no local resources in its business (such as computer production, airplane manufacture, or pharmaceutical formulation), then the company's bargaining position is relatively stronger.

The history of the international petroleum industry has shown quite clearly the power of host governments to control their mineral wealth. Since the early 1970s , foreign MNEs have been forced into positions of supplying technical expertise, overseas transportation, and other contractual assistance, without holding controlling ownership of oil reserves. In Mexico this nationalization took place in 1938, but in Venezuela it was 1975 and in Ecuador, 1974 and 1977. Similar results have occurred in the bauxite (primarily in Jamaica) and copper (in Peru and Chile) industries.

Sixth, the power of the government is greater in situations where the market served by the business is entirely in the host country. Again, since the government has sovereignty over its economy, *access to the domestic market* can be offered or restricted as the government chooses. This dimension is really the other side of hypothesis 2 (second part) above, in which export marketing was the focus. If export markets are unimportant, then the bargaining strength rests with the government, and vice vesa. At the present time, most manufacturing and service-sector investment in Latin America is used to serve the local market, while much of the raw materials investment goes to serve foreign demand. So the two ends of the spectrum can be categorized in broad terms as manufacturing and services (where the government has the advantage) and raw materials (where the MNE has the advantage). Unfortunately, it was not possible to explore this dichotomy with the available sample, since all but two of the firms were in manufacturing businesses.

The previous measure is complicated by the widely diverse nature of the businesses involved. Another way to measure the importance of the local market is to compare countries as to their size and restrictiveness toward MNEs. The larger countries should be able to impose more restrictions on MNEs seeking to enter their (more desirable) markets. Without devising a complex measure of this phenomenon, it can be asserted that, with the demise of Decision 24 of the Andean Pact (see Ch. 6 below), Brazil and Mexico do present a more restrictive legal environment than do the smaller countries of the region.

Seventh, in *high-visibility situations*, when dealings between the government and company are widely publicized in the press and other media, the government tends to have an advantage. That is, if foreign MNEs can be successfully portrayed as foreign interlopers, then the government can utilize public opinion to sway negotiations toward more favorable outcomes. This is especially likely to be true in raw materials' extraction, communications and public transportation, and public health.

These situations arise irregularly but consistently, with the government receiving the better result when it is pitted against a 'foreign' MNE. The tumultuous nationalizations of the copper companies in Chile during 1968–73; the reactions against ITT by many Latin American governments for its attempt to orchestrate the overthrow of the Allende government in Chile during approximately the same period; the reaction against United Fruit Company in Guatemala when its president was found to have given a US$1 million bribe to the host government; and many more examples can be listed to show how high-visibility disputes between government and MNE are invariably won by the government. (This does not necessarily mean that the government obtains the best results, since by penalizing the MNE, the government may lose that firm's supply of technology, access to foreign markets, etc.)

Eighth, the power of the government is greater in *highly competitive industries*, where more than two or three foreign firms are able to supply the product or service (such as tourism, production of standardized products, banking). In this situation, the government may be able to play the firms against each other to obtain the outcome most favorable to the country. By contrast, in industries or product lines where only one or two firms compete (such as in production of nuclear power plants or the latest generation of computer equipment), those firms can act more monopolistically and offer less concessions to the host government. This hypothesis was tested using the Fortune 500 data specifically to see if any difference in ownership percentage (100 percent vs joint venture) existed between firms with a two-firm concentration ratio of more than 70 percent for either of their two main products and firms with less. The chi-square test yielded a value of 0.2, which is insignificant. Evidently, using this data, industry competitiveness does not show a high correlation with the level of regulation of MNEs.

A ninth advantage that favors alternately the MNEs and the host governments is the *state of the Latin American economy*. During periods when there is a general recession in the region (e.g. the 1930s and the 1980s), bargains tend to favor the foreign multinationals. During periods when there is broad growth in the region, bargains tend to favor the host governments (e.g. the 1960s and 1970s).

Because of the long time period encompassed by this hypothesis and the relative lack of data on companies and regulations before World War II, the hypothesis will be explored anecdotally. Examining only the decades since World War II, we see a widespread openness toward foreign firms during approximately 1945–65, followed by a period of increasing restrictiveness during 1965–80, and then followed by another period of opening to foreign involvement during the 1980s. This general statement is illustrated by the Andean Pact's imposition of Decision 24 on foreign direct investment in 1971 and its subsequent reduction in the restrictions in the early 1980s, followed by full elimination of the rule in 1987. It is also illustrated by the attempts of Latin American governments to 'privatize' some of their government-owned companies in the mid-1980s, selling them to domestic and foreign investors.

A tenth and final advantage that favors alternately the MNE and the host government is the *obsolescing bargain*. In any bargaining situation, the initial negotiation for entry by the MNE favors the firm, that offers some attractive characteristics such as technology, market access, economies of scale, and/or other factors. Once the firm has sunk an investment into the host country, it becomes to some degree captive to the host government, that has sovereignty over business activity within its borders. This sequential process has been demonstrated time and again in government-company negotiations, particularly starkly in the raw materials industries that have been nationalized in the late 1960s and 1970s. Kobrin (1987) found that the obsolescing bargain was *not* a general characteristic of manufacturing investments in LDCs.

Conclusions

The bargaining theory as elaborated above undoubtedly offers insights into the relationship between host governments in Latin America and foreign MNEs that do business in the region. During the past decade, quite a few additional empirical studies in particular industries and countries have lent support to this perspective.

The difficult task that remains is to isolate the relative importance of the various bargaining factors that have been identified and to find additional important ones. To determine company strategy or government policy, it is necessary to understand the full complexity of these factors that coexist and to bargain accordingly. Proprietary technology or access to raw materials are not, by themselves, sufficient bargaining advantages to assure the possessor a sustained bargaining success. Much more work needs to be done to understand the interactions among factors and to suggest appropriate policy/strategy responses.

Notes

1. Actually, measuring a government's participation in the economy as its contribution to GDP may be an inadequate indicator for other reasons. For example, the government sector typically employs far more people as a percentage of the labor force than do private-sector. In Peru, the government sector employs almost one-half of the non-marginal workforce.
2. Wiarda, Howard, and Harvey Kline, *Latin American Politics and Development*, Boulder, Colo.: Westview Press, 1985, p. 76.
3. Moran, Theodore, *Multinational Corporations*, Lexington, Mass.: Lexington Books, 1985.
4. The Brazilian situation is discussed in more detail in Chapter 13. Those rules and the restrictions in Venezuela are also discussed in Business International Corporation's *Investing, Licensing, and Trading Conditions*, a notebook of country analyses that annually surveys laws and business conditions of interest to foreign firms.

Bibliography

Bennett, Douglas, and Kenneth Sharpe (1985) *Transnational Corporations versus the State*, Princeton, NJ: Princeton University Press.

Encarnation, Dennis and Louis T. Wells, Jr. (1985) 'Sovereignty En Tarde: negotiating with foreign investors', *International Organization* (Winter).

Fagre, Nathan, and Louis T. Wells, 'Bargaining power of multinationals and host governments,' *Journal of International Business Studies* (Fall).

Gereffi, Gary (1985) 'The renegotiation of dependency and the limits of state autonomy in Mexico (1975–82)' in Theodore Moran (ed.) *Multinational Corporations*, Lexington, Mass: Lexington Books.

Grosse, Robert (1988) 'Competitive advantages and multinational enterprises in Latin America', *Discussion Papers in International Business # 88-8*, University of Miami (September).

Kobrin, Steven, (1987) 'Testing the bargaining hypothesis in the manufacturing sector in developing countries', *International Organization* (Autumn).

Moran, Theodore, (1974) *Multinational Corporations and the Politics of Dependence: Copper in Chile*, Princeton, N.J.: Princeton University Press.

Moran, Theodore (ed.) (1985) *Multinational Corporations*, Lexington, Mass.: Lexington Books.

Wiarda, Howard, and Harvey Kline (1985) *Latin American Politics and Development*, Boulder, Colo.: Westview Press.

Chapter five

The economic impact of foreign direct investments: a case study of Venezuela

Introduction

Multinational enterprises (MNEs) are the single most important vehicle today for effecting international transfers of funds, technology, management skills, and products. The costs and benefits of these firms' activities in home and host countries are the subject of intense debate in the United States and abroad, as governments try to devise ways to harness MNEs' economic and political power.[1] The purpose of this chapter is to add to the understanding of *economic* impacts of multinational firms in host countries. Specifically, the chapter presents findings of a study of foreign manufacturing MNEs operating in Venezuela.

Efforts to measure the economic effects of foreign direct investment (FDI), or of MNEs generally, have been undertaken frequently during the past two decades. Governments of both industrial and less-developed countries sponsor such studies in an almost continuous stream. Some very useful findings have been published in a wide variety of contexts: the Reuber *et al.* study (1973) reporting on a large project in several less-developed countries, a similar study by Lall and Streeten (1977), and Caves' overview of the issues (1982, Ch. 9) are three of the best and most comprehensive ones. Conceptually, the goal is to evaluate the social costs of FDI projects and compare them with the social benefits — the results of this evaluation then must be compared with alternative sources of the funds, technology, market access, and/or other components of FDI that the host country is seeking.[2]

The economic impacts will be grouped into five categories. They have appeared often in the literature, and some examples of each are noted here. First, many analysts examine specifically the *employment* effects of FDI on the host country (e.g. International Labour Office 1981; Stewart 1976). Others focus on the ways in which MNEs affect *national income*, both its level and its distribution (e.g. US Tariff Commission 1973; Lall and Streeten 1977). A third group of analyses explore impact of foreign direct investment on a country's *balance of payments* (e.g.

Hufbauer and Adler 1968; Lall and Streeten 1977). Fourth, the impact of MNEs on the *transfer of technology* has been investigated (e.g. Biersteker (1978); Reuber *et al.* 1973). Fifth and finally, the impact of foreign investment on *industrial structure* in the host country has been studied (e.g. Lall 1978; Wells 1983). While this listing of the economic impacts of MNEs is by no means exhaustive — it completely ignores social and political issues, for example — it does cover all of the economic issues on which substantial literature exists. The present chapter attempts to uncover empirical evidence in each of these areas with respect to foreign MNEs that are manufacturing in Venezuela.

The next section explains the hypotheses and methodology of the study. The five sections following that present empirical findings concerning the impact of FDI on employment, national income, balance of payments, technology transfer, and industrial structure. The last section summarizes the results and offers some conclusions.

Methodology and hypotheses

Methodology

The initial approach in this analysis is microeconomic, that is it involves data-gathering at the level of the individual firm. While some data are available to measure *balance-of-payments* impacts and aggregate *national income* impacts at the macroeconomic level, most of the other issues only can be examined through company-level study. At the outset, it was decided to look only at manufacturing firms, thus ignoring MNEs in the extractive and service sectors. The impacts of each type of MNE tend to be somewhat different (see, for example, Caves 1982); and due to the constraints of time and money, this project was limited to the manufacturing sector.

A less-developed country was chosen as the host country to be studied, since the impact of FDI is most often cited in this context, and since the criticism of MNE activities has been most negative from LDCs. Venezuela was chosen specifically because it has a sufficiently large internal market to attract foreign manufacturing firms (i.e. it does not only attract offshore assembly or extractive industry.) Also, Venezuela has enough foreign firms currently operating such that a survey of reasonable size could be carried out. In addition, it was important to find a country with a number of domestic firms involved in the same industries, so that comparisons could be made between similar foreign and local firms. Some indications of recent FDI in Venezuela by US-based firms are presented in Table 5.1. Note that about 50 percent of US-based foreign direct investment in Venezuela is in manufacturing industries, with virtually no mining or petroleum investments (after the oil industry nationalizations in 1975).

Table 5.1 US direct investment position in Venezuela, by industry (US$m)

Year	Total	Mining	Petroleum	Manufacturing	Services
1966	2136	s	1544	281	190
1979	2241	78	1440	416	125
1975	2065	s	861	678	459
1980	1897	*	39	1035	522
1985	1548	*	133	837	578

Source: US Department of Commerce, *Survey of Current Business*, various August issues.
s = Suppressed to avoid disclosing individual company data.
* = Less than $500,000.

The sample of firms was intended to include all of the foreign MNEs and their main local competitors in a variety of manufacturing industries. An initial choice of industries was made by examining the Venezuelan-American Chamber of Commerce membership list. The Chamber has as members several hundred firms involved in international business, including the majority of large foreign-owned manufacturers. Foreign-owned Chamber members actually involved in manufacturing in Venezuela numbered ninety-seven. From this list, only about sixty have country headquarters in the capital and largest city, Caracas. The rest were eliminated from consideration, since the costs of carrying out more interviews in Valencia, Maracay, or elsewhere were judged prohibitive. The eight industries represented by this group are automobiles; chemicals; cosmetics; electrical appliances; food processing; pharmaceuticals; tires; and miscellaneous (mostly non-electrical equipment).[3]

Hypotheses

The main hypotheses examined in the study are as follows:

1. Foreign direct investment increases local employment in the investing industry and in ancillary industries.
2. Foreign direct investment increases the economic growth rate of the receiving country.
3. The overall balance of payments impact of foreign manufacturing investment is positive for the host country.
4. MNEs transfer more technology to the host country than that created or imported by local firms in the same industries.
5. MNEs have an overall positive impact on host-country industrial structure, because they create additional competition in the market.

Each hypothesis is presented in the positive direction: that is, if the evidence supports the hypothesis, then the impact of MNE activities will be viewed as positive from the host-country perspective. Since none of the main hypotheses was testable in the form shown above, each one was divided into parts and variants that focus on specific aspects of the central point. They are discussed in the following five sections.

Employment effects

Direct and indirect employment effects of foreign direct investment in host countries include jobs provided by the MNEs, jobs generated in local supplier and customer industries, jobs created due to multiplier effects of the MNEs' local expenditures and taxpaying, and jobs lost due to the new competition. Figure 5.1 suggests the range of employment effects.

Job creation and other related employment effects of foreign MNE activities that were examined here can be grouped under four headings: the capital-labor ratio, wages and salaries; use of expatriates; and net addition of jobs.

Capital-labor ratio

It is argued that MNEs from industrial countries are too capital-intensive for the local, high-unemployment environments of most LDCs. Mason (1973) and Stewart (1976) both found this to be true; that is, that foreign MNEs operating in the Philippines and Mexico and in Ireland respectively, did indeed have greater capital–labor ratios than their domestic counterparts. (From another perspective, MNEs have been found to use substantially more labor-intensive technology in their LDC affiliates than in *their own* industrial-country operations (cf. Morley and Smith 1977; Courtney and Leipziger 1975). The results of the present sample of firms in Venezuela, are shown in Table 5.2a. There is no statistically significant difference in the capital–labor ratios of Venezuelan versus foreign firms; this finding was reinforced by managers' perceptions of capital intensity. (The managers response to the question, 'is your company more mechanized; that is, do you use more machinery

Table 5.2a Capital intensity of foreign and Venezuelan firms

	*Average capital–labor ratio**
Venezuelan	291,783 bolivars/employee
Foreign	232,388 bolivars/employee

* Ratio defined as value of assets divided by number of employees.
Student's *t* for testing equivalence of means = 0.855, which implies a significance of 0.399.

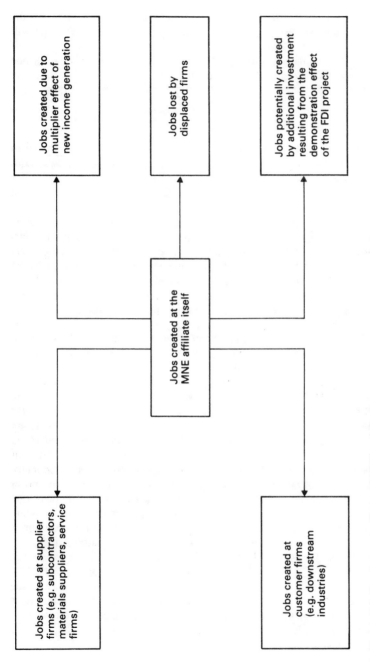

Figure 5.1 Employment effects of FDI projects

and less workers, than your competitors in Venezuela?') These results are shown in Table 5.2b.

Table 5.2b Capital intensity of foreign and Venezuelan firms: managers' responses to the capital-intensity question

	Greater capital intensity	Equal capital intensity	Less capital intensity
Venezuelan	4 (44%)	4 (44%)	1 (11%)
Foreign	11 (39%)	16 (57%)	1 (4%)

$x^2 = 0.96$, x^2 crit 2,0.05 = 5.99. Thus no significant difference between response frequencies for Venezuelan vs foreign firms was detected.

These findings in Table 5.2b contradict the previous studies by demonstrating that foreign firms in this sample do *not* have greater capital intensity than their local counterparts. The reason for such a result may be that the firms are of similar age, with an average start-up date of 1955 for both Venezuelan and foreign firms and thus thirty years of adaptation to the Venezuelan conditions. Also, the fact that foreign technology provides the base for the manufacturing process in each of the firms means that capital intensity for the sampled Venezuelan firms may be higher than for Venezuelan firms in other industries using greater local technology. No more certain reason for this surprising finding could be deduced from the evidence.

Wages and salaries

Previous studies have consistently shown that foreign multinationals pay higher wages and salaries than do local firms in the same industries (e.g. Mason 1973; Stewart 1976; and Ingles and Fairchild 1978). This phenomenon may be due to the fact that MNEs must offer higher compensation to attract qualified people away from their existing jobs and/or to the fact that MNEs employ some expatriates, whose compensation reflect premia for the overseas assignment. Our findings are shown in Table 5.3. These results fail to confirm previous studies, since there is no significant difference between the compensation patterns of foreign MNEs and local firms. Unfortunately, without specific salary and wage data, which were unavailable to the interviewers, clear conclusions cannot be drawn here. Once again, the similarities in compensation may be due to the similarity in capital–labor ratios between local and foreign firms (or they may be due to a reporting bias that leads managers to rank their own firms as higher- or equal-paying, but not lower-paying in their industries.)

Table 5.3 Salaries and wages

		Salaries			Wages	
	Higher	*Equal*	*Lower*	*Higher*	*Equal*	*Lower*
Venezuelan	5 (50%	5 (50%)	0	4 (40%)	6 (60%)	0
Foreign	10 (34%)	18 (62%)	1 (3%)	13 (45%)	15 (52%)	1 (3%)

$x^2 = 0.994$ for salaries and 0.479 for wages. x^2 crit $2, 0.05 = 5.99$. There is no significant difference detected between Venezuelan and foreign company compensation policies as measured.

Use of expatriates

On the surface, it may appear that the greater use of non-local personnel by MNEs compared with local firms means that the MNEs are simply providing fewer jobs to locals by importing some of their executives and technical staff. In fact, the reality is considerably more complex. For example, expatriates may be used on a temporary basis to train locals for the same positions. Expatriates also may be used when the firm needs skills that are not available locally and when the firm cannot afford to train a local person to perform the particular task. In addition, executives may be assigned for short periods to different countries to build their experience and knowledge of the countries — which necessitates using expatriates of various nationalities in a variety of countries. Finally, the multinational firm may choose to exercise control over a foreign affiliate by placing an experienced and trusted expatriate in charge, thus implicitly pushing the affiliate to follow a path that the home office will support. For all of these reasons, MNEs can be expected to use more non-local personnel than local firms; and our findings weakly bear out this expectation (Table 5.4). The number of expatriates employed per firm (namely 2.9 for Venezuelan versus 4.4 for foreign firms) is not significantly greater for the MNEs than for Venezuelan firms. (Student's *t* for expatriates/firm, comparing Venezuelan and foreign firms, = 0.7379, which implies a significance level of about 0.45.)

Table 5.4 Use of expatriate personnel by Venezuelan and foreign firms

Type of firm	Total expatriates*	US expatriates	Latin American expatriates	Total employees
Venezuelan	29	2	0	7,172
Foreign	127	32	56	20,637

* Subcategories do not add to total because some respondents did not identify the source of the expatriates.
Student's *t* comparing expatriates/employees for Venezuelan versus foreign firms = 0.3156, which implies a significance level of approximately 0.75.

Net addition to jobs

This issue is the most complex of those related to employment in the study. My intention is to measure the net addition (or reduction) to jobs in Venezuela due to the MNEs' activities there. This *net* addition requires some estimate of the employment that would have existed had the MNEs not invested; and such an estimation requires assumptions as to whether or not a local firm would have undertaken a similar investment project if the MNE had not entered. Some authors have created alternative scenarios to explore the impacts under assumptions of (1) total replacement of FDI, (2) no local investment.[4] In the present context, we will simply examine some of the specific ways in which local employment is augmented. One reason to expect that MNEs add to local employment is that most of the foreign firms established new companies when they entered Venezuela (rather than simply acquiring existing firms), and most of their hiring has been done from local sources. Table 5.5 shows the means of entry into Venezuela. New companies (or new joint ventures) imply new jobs, especially when hiring is known to be done locally, as shown in Table 5.6a. Although MNEs did transfer a handful of employees into Venezuela from other affiliates elsewhere, they hired over 99% of their managerial and technical staffs and virtually 100% of their non-salaried employees locally. (Total expatriates relative to total local employees for the foreign multinational firms in the sample was: 127/20637 = 0.00615, or less than 1 percent.)

These data cannot fully substantiate the claim that MNEs add to host-country employment, but all indicators show positive evidence that this is indeed the case.

Table 5.5 Entry into the Venezuelan market

Type of firm	Acquire existing firm	Set up new company	Set up a joint venture
Venezuelan	1	7	1
Foreign	6	18	5

$x^2 = 0.763$, which fails to reject the hypothesis of a difference between frequencies of Venezuelan and foreign firms' entry strategies.

Table 5.6a Source of hiring: technical staff and managers

Type of firm	Other affiliates of our firm	Local firms	Local universities	US universities	Other MNEs
Venezuelan	3 (14%)	6 (29%)	7 (33%)	1 (5%)	4 (19%)
Foreign	17 (18%)	22 (23%)	24 (25%)	14 (15%)	19 (20%)

Table 5.6b Sources of hiring: workers

Type of firm	Other affiliates of our firm	Local firms	Local schools	Other local sources
Venezuelan	2 (13%)	6 (38%)	4 (25%)	4 (25%)
Foreign	6 (11%)	18 (34%)	11 (21%)	18 (34%)

Income effects

Measures of income effects can include estimates of the increase in the *amount* of national income generated by FDI to evaluations of income *distribution* between capital and labor or between other income groups. This study examined several measures of impact on national income and did not consider directly the issue of income distribution.[5]

One piece of information regarding distribution is worth noting. The normal presumption in terms of *functional* distribution of income is that FDI, if it does add to local capital formation, tends to reduce the return to local capital (by increasing the supply of capital) and raise the return to local labor. Judging from the wage and salary information presented above, part of this reasoning appears to be true in the Venezuelan case — although it was not possible to ascertain if the return on capital indeed fell due to FDI.

Looking then at income growth, we dealt with four sub-hypotheses relating to: investment substitution, reinvestment of earnings, tax payments, and demand patterns.

Investment substitution

As noted previously, the hypothesis that FDI substitutes for local investment that would have occurred otherwise is most difficult to demonstrate. One cannot show counterfactually what would have happened if the MNEs had not set up their factories, plants, offices, and so on, in the host country. However, some indications as to how much of FDI represents additional investment (and how much substituting investment) do exist. For example, the measure in Table 5.5 showed that the vast majority of FDI went into new companies or joint ventures (79 percent), which may add to local investment, rather than acquisition of existing, operating firms, which would replace existing investment. Also, the measure in Table 5.7 shows the percentage of firms that undertook local production to supply the Venezuelan market with products that were formerly imported; this, too, points to net income gains. Clearly, FDI substantially replaced formerly-imported products, which implies that it added local production and reduced sales of imported products in the local market. (Investment by local firms also replaced imports in half of the cases examined, so one cannot argue that MNEs are necessary to carry out the import replacement.)

Table 5.7 Percentage of firms producing products that formerly were imported

Type of firm	Formerly imported	Not formerly imported
Venezuelan	50%	50%
Foreign	83%	17%

$x^2 = 24.44$ and x^2 crit 1,0.05 = 3.84. This implies that there is a greater propensity for foreign firms to operate import-replacing businesses than for Venezuelan firms.

Another way to consider the contribution of FDI to raising national income would be to use macroeconomic data and estimate the determinants of national income without including FDI, and then see if FDI offers any additional explanatory power when added to the equation. A simple accelerator model of investment is used to estimate gross domestic product:

GDP = a + b1 (Domestic Investment) + b2 (GDP in previous year).

Using data for the period 1970–1985 resulted in these estimates:

GDP = 1.50 +   1.32(DI) +   0.65(GDP-1).   [$R^2 = 0.95$]
       (0.49)   (7.91)   (4.02)

Adding foreign direct investment to the equation yields these results:

GDP = 1.54 +   1.32(DI) +   0.66(GDP-1) + 1.48(FDI).   [$R^2 = 0.95$]
       (0.46)   (3.83)   (7.28)   (0.33)

T-statistics for the coefficients are shown in parentheses below the equations. Coefficients of multiple correlation in both cases exceeded 0.90, primarily due to the explanatory power of lagged GDP. Both lagged GDP and domestic fixed capital formation (DI) are highly significant in the expected direction. Foreign direct investment has the expected sign, but it is not significant in the aggregate form used. Using a second measure of FDI that more precisely covers the kinds of firms in the sample, that is US *manufacturing* investment in Venezuela, the revised equation showed these results:

GDP = -1.04 +   1.02(DI) + 0.64(GDP-1)   1.03(FDI). [R = 0.96]
       -(0.19)   (1.62)   (7.13)   (1.66)

In this case, both investment terms only are significant at the 80 percent confidence level, but the fit of the regression is marginally better, and less serial autocorrelation is evident in the error terms. The total regression equation provides weak support for the hypothesis that FDI contributes to increasing host-country GDP.

Reinvestment of earnings

The hypothesis in this case is that foreign MNEs reinvest less of their earnings than comparable local firms. Such a view stands to reason, since shareholders of the MNE are in a foreign country, and typically some of the earnings of subsidiaries are distributed to them. On the other hand, it may be that Venezuelan shareholders of the local firms actually take their distributed earnings and invest some or most of them abroad — especially in times of economic crisis and low confidence in the country, as have existed during most of the 1980s in Venezuela. None the less, the direct results of earnings' distribution by MNEs versus local firms are that MNEs send profits abroad and domestic firms distribute them locally. Our measure of the simple earnings distribution is as shown in Table 5.8. Virtually all of the firms reinvest all or most of their earnings in the company. No effort was made to try to estimate the amount of Venezuelan earnings that were subsequently invested abroad rather than reinvested within the country.

Table 5.8 Profit remittances to parent firms (as percentage of total profits)

Type of firm	0%	1-25%	26-49%	50% or more
Venezuelan	3	1	0	0
Foreign	11	12	4	2

Note: The parent company for a Venezuelan firm typically was the Venezuelan holding company that owned the firm in the sample.

Table 5.9 presents aggregate data reported by the US Department of Commerce for use of earnings by all US firms' affiliates in Venezuela. Note that, according to the US Department of Commerce data, profit remittances constitute almost half of earnings of the Venezuelan affiliates. This definition includes interest payments on intra-company accounts and earnings of unincorporated branches as remittances, so it overstates the dividends paid by the subsidiaries in the present sample. More importantly, the firms in the present sample were responding on the basis of profits *including* 1983, when losses precluded any remittances. The discrepancy between survey responses and Commerce Department data, therefore, is more apparent than real.

Table 5.9 Uses of income from US-owned manufacturing companies in Venezuela

1980–5 annual averages*	Interest & dividend remittances	Retained earnings	Total income
Current dollars	$US46 million	$US61 million	$US107 million
Percentages	(43%)	(57%)	(100%)

Source: US Department of Commerce, *Survey of Current Business*, August issues, 1981–6.
* These averages *exclude* 1983, when the economic crisis caused US firms to declare massive losses that outweighed the gains of the other four years.

Tax payments

Given their ability to move funds and other resources between countries, MNEs would be expected to adjust transfer prices, fee schedules, and other intracompany payments to reduce global taxes. MNEs presumably would try to show low profits in high-tax jurisdictions, and show high profits where less is paid in taxes. Additionally, such firms can be expected to try to move funds out of risky locations, for example, where sudden devaluation or currency controls are likely. Considering taxes, Venezuela should be a location of low MNE profits, since its corporate tax rates are higher than in the United States and many European countries. (The highest tax bracket has a rate of 50 percent.) Considering risk, Venezuela probably would be viewed by MNE financial managers as a relatively risky environment, and thus one where low profits would mean less financial risk. Because the two influences lead in the same direction, our hypothesis would be that MNEs should show lower profitablility and subsequent tax payment relative to local firms.

The hypothesis is more difficult to state, however, since other important influences affect the profitability of MNEs in Venezuela. For example, based on their technological and marketing advantages, MNEs are expected to earn higher profits than local competitors. The net result is that we will approach the issue with no a priori expectation, and simply test to see if MNE profitability, and thus tax liability, is similar to that of local competitors. The values of taxes paid relative to company sales are shown in Table 5.10. Since their sales (and thus, presumably profits) are larger, on average, the foreign MNEs paid somewhat more Venezuelan corporate income tax than their local counterparts. In relative terms, adjusting for the difference in firm sizes by measuring tax payments divided by sales, the foreign MNEs also experienced a higher tax burden than the Venezuelan firms. Although the failings of both measures are serious, both do offer evidence that MNEs are more profitable then their local competitors, and they therefore pay more Venezuelan tax.[6]

Table 5.10 Taxes paid by MNEs and local competitor firms

Type of firm	Tax paid per firm	Tax/Sales per firm
Venezuelan	$3.75 million	4.44%
Foreign	$4.40 million	5.09%

Demand patterns

MNEs may affect host-country demand patterns by using greater (and perhaps more effective) promotion than local firms. This proposition was examined only in light of the advertising expenditures of local and foreign firms. Table 5.11 compares advertising expenditures as a percentage of company sales for foreign and Venezuelan firms. Foreign firms do indeed demonstrate a significantly higher propensity to advertise than local firms in the sample. From the information collected, it is not possible to conclude that national demand patterns were altered by MNE promotion efforts, since no measure was taken of promotion in other industries. However, within the industries studied, MNEs do place more resources into this aspect of marketing than their local competitors.

Table 5.11 Advertising expenditures as a percentage of sales

Type of firm	None	1-2%	3-4%	5% or more
Venezuelan	3	2	2	2
Foreign	1	8	10	9

$x^2 = 6.29$, and x^2 crit 3,0.05 = 7.82. The difference in advertising intensity is significant only at the .10 level (since x^2 crit 3,0.10 = 6.25.)

Whether or not the MNEs affect national demand patterns, this issue strictly lies outside of the realm of economic cost-benefit analysis. That is, consumer preferences, which may be 'genuine' or 'acquired', cannot be evaluated with this type of economic analysis.[7]

Balance-of-payments effects

The balance-of-payments effects of FDI are far more complex than the initial inflow of capital from abroad and the subsequent stream of profit remittances to the parent company. To the extent that FDI (and then local production) often replaces imports, it should be considered as lowering the host country's imports. (Sometimes imports of equipment and parts, as well as additional product lines, replace most or all of the value of products originally imported.) Also, some manufacturing FDI leads to exports to unrelated customers or to the MNE's own affiliates elsewhere; this certainly adds to the favorable balance-of-payments effect of the project. Ultimately, of course, the goal is to compare the balance-of-payments effects of an FDI project with those of the other alternatives available to the country for obtaining the same products or services. Figure 5.2 depicts the range of balance of payments effects related to FDI for the host country.

In the present study, four specific measures of balance-of-payments impacts were examined: substitution for imports; purchase of inputs from abroad; exports from Venezuela; and profit remittance and other financial transfers abroad.

FDI substitutes for imports

This hypothesis has been tested previously by Lall and Streeten (1977) and Adler and Stevens (1974), among others, who found that manufacturing FDI does indeed substitute for imports. The evidence to support this hypothesis in the present study already has been shown in Table 5.7. Of all the MNEs, 83 percent used FDI to replace products that they previously had imported into Venezuela — and even half of the Venezuelan firms reported that they began producing locally after previously importing their products. The precise amount of import substitution is not possible to judge, since some of the FDI may have occurred because previous imports were becoming uncompetitive, and they necessarily had to be produced locally (by MNEs or local firms) in order to remain in the market.

Purchase of inputs from abroad

While FDI substitutes for imports to some degree, the direct investment project itself may result in additional imports of equipment, materials, and other products. The US Tariff Commission study (1973) found that such additional importing was common among US-based foreign investors. In the present study, purchase of factor inputs by MNEs was compared with purchase of imported inputs by local competitors. Tables 5.12 and 5.13 show the results. The MNEs tend to purchase from more local suppliers than do local competitors in the sample, although the difference is not significant. Another view of the relative use of local suppliers can be obtained by examining their contribution to the sample firms' total costs. The results of this comparison are shown in Table 5.11b and are rather surprising, since MNEs were expected to have substantially greater links to foreign suppliers (e.g. their own affiliates in other countries)[8] — but import-dependence is not significantly different between Venezuelan and foreign firms. This may be due

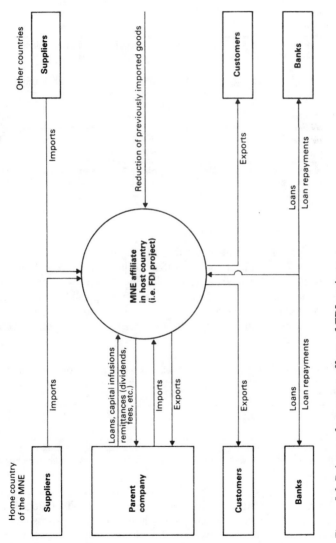

Figure 5.2 Balance-of-payments effects of FDI projects

partially to government policies that encourage or require local purchasing of inputs, and to the long experience of most of the foreign firms in Venezuela (and their consequent ability to find acceptable local suppliers).

Table 5.12 Number of local suppliers used for production inputs

Type of firm	0-20	21-50	51-75	76-100
Venezuelan	2	2	3	2
Foreign	1	7	10	11

$x^2 = 3.58$, and x^2 crit 3, 0.05 = 7.82. This implies that there is no significant difference in propensity to use local suppliers for either group of firms.

Table 5.13 Percentages of total costs attributed to local and foreign suppliers

Type of firm	Local suppliers	Foreign suppliers
Venezuelan	56.4%	43.6%
Foreign	54.8%	44.8%

$x^2 = 0.04$, and x^2 crit 1, 0.05 = 3.84. This implies no significant difference between Venezuelan and foreign firms' use of local relative to foreign suppliers.

Exports from Venezuela

Previous studies have found that MNEs tend to export more of their production from host countries than do local firms (e.g. Mason 1973; Hunt 1972). This appears reasonable, since MNEs have foreign affiliates and knowledge of foreign markets that may provide more outlets for local production than those accessible to local, non-multinational firms. In the present study, very little exporting was done, as shown below in Table 5.14.

Table 5.14 Exports from Venezuela (as a percentage of total sales)

Type of firm	None	1-25%	25-49%	50-99%	100%
Venezuelan	3	0	6	0	0
Foreign	23	0	6	0	1

Ignoring the columns with zero entries, $x^2 = 7.15$, and x^2 crit 2, 0.05 = 5.99. This implies that the Venezuelan firms have a greater propensity to export than the foreign firms.

The result is consistent with other findings about MNE strategies in Latin America (e.g. Grosse 1983), in which manufacturing firms were found to invest almost always to serve the local market. Transportation costs and legal barriers make export sales uncompetitive in other countries. Apparently, in this region the MNEs are not able to take advantage of many inter-country shipments of their products. (On a global scale, manufactures from protected Latin American economies tend to be priced out of other markets.) Interestingly, the local firms in the sample did demonstrate a relatively large propensity to export, in comparison.

Financial transfers to the parent company

Profit remittance, royalties and fees, and intra-company loans constitute another area of balance-of-payments impact of MNEs. It was expected that these firms would transfer some or all of their earnings to the parent company, and that other financial transfers would be used to pay for technical and managerial assistance (namely royalties and fees). Vaitsos (1974) and others have argued that these transfers often outweigh the value of the initial investment by the MNE. (While these transfers are an inadequate measure of the *total* balance-of-payments impact, they still are quite important.) Tables 5.8 and 5.9 showed data on profit remittances, and the discussion below focuses on other types of transfer.

Royalties and fees are expected to be minimal from Venezuelan affiliates of foreign MNEs, since local law prohibits such payments. Many exceptions are granted, in fact, but the level of royalties and fees is quite low. All the firms in the sample reported paying *no* royalties or management fees to their parent offices. Aggregate data reported by the US Department of Commerce for 1983 showed a value of $7 million in royalties and fees paid from Venezuelan affiliates of US parent companies (in all industries); this compares with $129 million of profit remittance for the same year.

Financing of the Venezuelan operation is another potential area for balance-of-payments impact. Local subsidiaries can be financed through retained earnings, new investment by the parent, and local or foreign borrowing. Table 5.15 shows how the sample firms finance their operations. Of the financing methods chosen, the vast majority of firms use local sources of funds, that is bank borrowing and retained earnings.

Table 5.15 Financing of Venezuelan operations

Type of firm	Retained* earnings	Local borrowing	Parent company	Foreign bank borrowing
Venezuelan	4	9	0	1
Foreign	18	22	12	2

* These categories are *not* mutually exclusive.
$x^2 = 4.84$, and x^2 crit 3, 0.05 = 7.82. This implies no significant difference between financing methods used by Venezuelan and foreign firms.

A dozen of the MNEs stated that credit from the parent company is used as one method of financing, though all of them reported that local financing was most important.

Technology transfer

The issue of technology transfer is so multi-faceted as to preclude a single view of MNE impact. An aspect that often is criticized by LDC governments is the 'appropriateness' of the technology used by MNEs, which typically come from relatively capital-abundant countries and which utilize relatively labor-saving technologies. On the other hand, MNEs constitute educational centers for local managers and employees; the MNEs train people at the firm's expense and transfer skills to them which can be used in other applications in the country. Another concern of host countries is the location of R & D carried out by MNEs — the firms usually prefer to centralize large-scale R & D in the home country, while undertaking only marketing research and/or product development in host countries. A final concern to be examined here is the price charged by a parent firm for the transfer of technology to its subsidiaries; host-country governments often argue that it is too high.

Appropriate technology

Technology that is appropriate from the view of the host-country government often is not appropriate from the view of the firm. When a given technology is transferred to a new environment, some adaptation may be needed for the firm to use it most efficiently. The receiving country also may prefer to have the technology altered to utilize its productive factors in different degrees. These two types of adaptation are *not* equivalent. When an MNE undertakes a horizontal investment to manufacture an existing product in a new country, the firm typically tries to construct a facility similar to the original one. Adaptation occurs to adjust to different electric power, different climate, perhaps a different scale of output — but not to any major difference in capital or labor intensity. Were adaptation of the facility to take place such that labor intensity were raised, the cost would have to be borne by someone. Host governments that desire such adaptations need to formulate policy to account for these costs (and risks), if they want to attract specifically this type of FDI. Such more labor-intensive technology would be more 'appropriate' in a high-unemployment country, but the cost of creating this technology is far from zero.

This issue was pursued in the study first by comparing capital intensity of the MNEs versus local firms. Table 5.2 showed virtually no difference in capital intensity of Venezuelan and foreign firms in the sample. A second measure was used to explore the firms' efforts to adapt technology to the local environment. (Both local firms and MNEs often needed to adapt the available technology, because most of it came from industrial countries, especially the United States.) Table 5.16 compares technology adaptation by Venezuelan and foreign firms. More than one-quarter of the foreign firms responded that they adapted their 'home country' technology substantially when operating in Venezuela. Almost all of them noted that this adaptation made production more labor-intensive. This result is not significantly different from the technology adaptation used by local firms in the sample.

Table 5.16 Adaptation of production technology by MNEs and local firms

Type of firm	Unchanged from original use	Adapted substantially	Changed to more labor-intensive
Venezuelan	5	3	0
Foreign	16	6	5

$x^2 = 2.05$, and x^2 crit 2, 0.05 = 5.99. This implies that there is no significant difference in adaptation by Venezuelan versus foreign firms in the sample.

Skill transfers

Regardless of the capital- or labor-intensity of a host-country FDI project, the managers and workers hired to staff it will learn skills needed to operate it and perhaps to use in other applications in the host country. In addition to normal on-the-job learning, many firms offer training programs for their employees. The hypothesis in this case was that MNEs offer greater training opportunities than local firms. Tables 5.17 and 5.18 present results of this inquiry. At the management level, foreign MNEs do indeed offer more training to their personnel than local firms. In many cases this training includes short courses at the home office or other affiliates of the MNE — a much broader experience than what local firms can offer (Table 5.17). In the case of non-management training, both local firms and MNEs have programs in virtually all of the firms in the sample. Only one exception existed to this rule (Table 5.18).

Table 5.17 In-house management development programs

Type of firm	Training offered	Training not offered
Venezuelan	44%	56%
Foreign	87%	13%

$x^2 = 40.91$, and x^2 crit 1, 0.05 = 3.84. This implies a significant difference in use of management training programs between Venezuelan and foreign firms.

Table 5.18 In-house staff and worker development programs

Type of firm	Training offered	Training not offered
Venezuelan	100%	0%
Foreign	89%	11%

Location of R & D

Host governments generally prefer to see firms carry out R & D locally, so that the benefits may accrue to the local economy. MNEs, on the other hand, generally prefer to centralize basic R & D and much of their product development, because of the costs involved. While very little basic research is carried out by MNEs in LDCs, increasing amounts of marketing research and product development (adaptation) are being undertaken there. Table 5.19 compares the firms in the sample in terms of their R & D activity in Venezuela. There is significantly more R & D in Venezuela by the MNE affiliates than by their local competitors, supporting a claim that MNEs generate more technology development than would occur in their absence.

Table 5.19 Research and development expenditures in Venezuela (R & D costs as a % of sales)

Type of firm	None	1-2%	3% or more
Venezuelan	7	2	0
Foreign	7	17	6

$x^2 = 9.20$, and x^2 crit 2, 0.05 = 5.99. This implies a significant difference in R & D intensity between Venezuelan and foreign firms in the sample.

An additional measure of the technology transferred to Venezuela by the MNE affiliates is the sale of technology to local firms. Such sales appear to be more common by MNEs than by local firms, but still they are relatively infrequent (occurring in only four cases of the MNEs), as shown in Table 5.20. These technology sales add to the total gains to Venezuela, assuming that the technology would not have been obtained without the local presence of MNE affiliates.

Table 5.20 Technology sales to local firms

Type of firm	Do sell technology, e.g. license local firms	Do not sell technology, e.g. license local firms
Venezuelan	0%	100%
Foreign	13%	87%

The cost of technology transfer

The cost of transferring technology is not trivial, either for the firm or for the receiving country. Teece (1976) estimated that the cost to the *firm* for transferring technology from one affiliate to another is about 19 percent of the total cost of setting up the new affiliate. As for the cost to Venezuela, several aspects are important. First, Venezuelan law prohibits payment for technology transfer between a local MNE affiliate and the parent firm; however, some exceptions are made, and a relatively small amount of royalty payments is recorded each year by the US Department of Commerce from US-owned affiliates in Venezuela. Second, the product, process, and managerial knowledge that enters Venezuela through ongoing operations of MNE affiliates is not priced, but it clearly constitutes an inflow of technology. Sale of technology by MNEs to local firms presumably occurs on a commercial basis, so the cost there reflects the market's valuation of that technology (although the local presence of the MNEs makes such technology more available than from foreign sources). Internal transfer pricing of products by the MNEs provides another outlet for payment related to technology transfer. Indeed, Vaitsos (1974) criticizes multinational pharmaceutical companies for overpricing the products they ship to affiliates in Colombia. Numerous other authors have examined the issue in detail, though not with respect to Venezuela. The transfer pricing issue is one that LDC governments examine carefully, and about which MNEs are quite sensitive. No new evidence on this issue was collected in the present study.

Industrial structure

The final area of economic impact relating to FDI in Venezuela is industrial structure. How does FDI affect industrial concentration and the distribution of industry within the local economy? We explore four points under this heading: firm size; industrial concentration; relative profitability; and industry choice.

Firm size

Other studies have found that MNEs tend to be larger than their local competitors (e.g. Wells 1983; Ingles and Fairchild 1978). Evidence collected in this study supports this claim. The average annual sales of MNE affiliates in the sample was about $95.2 million per year, while the Venezuelan counterparts had sales of about $63.6 million per year. Larger size tends to permit greater economies of scale and generally a more favorable position on the experience curve, which enables the firm to lower costs more rapidly than competitors.

Industry concentration

It is not at all clear that the entry of MNEs has led to greater concentration of Venezuelan industry. Indeed, LDC markets often are characterized by small numbers of large local firms, protected by barriers against imports and other forms of government assistance. Entry of foreign competitors tends to erode this pattern of concentration. Since almost all of the FDI entered through establishment of new firms (rather than through acquisition of existing firms), it could be expected that concentration either declined or remained about the same.

Relative profitability

It already has been shown above and in Table 5.9 that MNE affiliates had greater sales and paid more taxes than their local competitors in the 1983–4 period of the study. One may infer from those data that MNE affiliate profits did indeed exceed those of the local firms. Table 5.21 presents another measure of profitability: the company managers' judgments about their relative profitability. This highly subjective measure of profitability shows that foreign company managers saw a slightly higher profitability of their affiliates than the Venezuelan firms, but the difference is not significant. Using data from the US Department of Commerce, the average profitability of manufacturing affiliates of US-based firms in Venezuela during the 1980s has been negative, due to the huge losses in 1983. Ignoring 1983, net income averaged about 9.1 percent of direct investment book values in manufacturing. This is slightly higher than average returns in Venezuelan manufacturing overall during the period (*and* it does not include additional MNE profits earned from fee payments to the home office and from intra-company shipments.)

Table 5.21 Profit comparison of Venezuelan and foreign firms

Type of firm	Higher	Same	Lower
Venezuelan	3	4	2
Foreign	12	15	1

$x^2 = 3.18$, and x^2 crit 2, 0.05 = 5.99. This implies no significant difference between profitability of Venezuelan and foreign firms in the sample.

It is unclear what conclusions can be drawn from these observations. Perhaps the MNEs are somewhat more efficient than their local competitors. Perhaps the local firms command lower prices because of lesser name recognition. In any event, the MNE affiliates do appear to be more profitable than the local firms.[9]

Industry choice

Previous studies have found that MNEs tend to flock to higher-profit and higher-growth industries (e.g. Ingles and Fairchild 1978). Because much of Venezuelan industry is privately owned, rather than having shares traded on a stock exchange, little evidence is available on relative industry profitability. Casual observation shows that MNEs tend to enter large market segments, and often high-price, high-quality segments in particular. Such segments in the United States are characterized by relatively high profit margins, in comparison with more standardized product segments.

Conclusions

The study of economic impacts of multinational enterprises in host countries has no widely-accepted methodology at present. Tools of cost-benefit analysis can be applied to foreign investment projects, but the range of benefits and costs to be measured remains the difficult problem.

The basic analytical principle is to evaluate social costs of FDI projects and compare them with social benefits, and then compare these results with alternative means of obtaining the desired investment, technology, employment, and so on. This paper argues that a useful method to handle the disparate types of costs and benefits is to examine each one separately; policy-makers may then choose weights for each goal. This reasoning resulted in an objective function for the present paper consisting of FDI impacts for the host government to optimize: (1) addition to employment; (2) addition to national income; (3) balance-of-payments impact; (4) transfer of technology; and (5) impact on industrial structure.

Only the manufacturing sector was examined in the Venezuelan sample of firms. This deliberate selection avoids the complications of mixing industries whose impacts can be expected to be very different, for example extractive activities and service industries. Manufacturing by multinational enterprises in LDCs generally has been found to substitute for imports, except for assembly operations that are used for export to industrial markets. The sample of firms in Venezuela was not chosen to emphasize either of these types in particular, but in fact there is very little offshore assembly in Venezuela, and most foreign-owned manufacturing exists to serve the domestic Venezuelan market.

Our findings in the five categories of costs and benefits were broadly positive in the sense that FDI appears to offer net benefits to the host country that either would not be available from other sources, or would be more costly to obtain through alternative vehicles. Conclusions must be tentative, however, because of the extreme difficulty of positing counterfactual alternatives to FDI, and also because of the difficulty of obtaining adequate data to perform the comparisons.

Subsidiaries of foreign multinational firms in the sample do not have greater *capital intensity* than their local competitors; they do remit more funds abroad than do local competitors; but they also transfer in more technology, provide more jobs, and develop more linkages to local supplier firms than the local competitors. The net result is quite difficult to judge without crucial assumptions — such as the amount of investment that would have occurred from local firms if the FDI projects had not taken place.

This study adds another set of empirical findings to the continuing effort to evaluate and deal with MNE activities in host countries. In addition, the conceptual structure may offer a methodology that can serve objective decision-making in Venezuela and other LDC host countries.

Notes

Acknowledgments: Sincere thanks go to Simonetta Franchi de Blohm and Marianela Hernandez, whose hard work in carrying out interviews enabled this project to be undertaken. Thanks also to the thirty-nine company managers who took their time and effort to participate in the project. Any errors of fact or interpretation remain the responsibility of the author.

*An abridged version of this article was published in *Management International Review* (4/1988)

1. See, for example, the US Tariff Commission report (1973); and the United Nations reports (1978, 1983).
2. Measurement techniques are discussed in Vernon and Wells, *Manager in the International Economy*, 5th edn, Englewood Cliffs, NJ: Prentice-Hall, 1986, Ch. 5.
3. Each company was contacted by telephone to request an on-site interview. For various reasons (from unwillingness to participate and unavailability of executives to unlisted phone numbers) only twenty-nine complete interviews with these firms were accomplished during the period January 1984 to December 1985. In addition, ten interviews with Venezuelan firms of similar sizes that were found to be important competitors in the same businesses were carried out during the same time frame. In total, thirty-nine firms were visited, including the leading foreign and Venezuelan firms in each of the eight industrial categories. In most cases the general manager or president was interviewed, and between one and three hours were spent per interview. Most of the interviews were done in Spanish, either by the principal investigator or by one of the two graduate students participating in the project.
4. See Hufbauer and Adler (1968).
5. Distributional effects may in fact be most important in the discussion of foreign participation in the domestic economy. However, in the present context it was impossible to pursue that subject.
6. A regression of tax payments on sales and nationality of firm shows an insignificant impact of nationality on taxes, though the explanatory power of the regression is quite low ($R^2 = 0.10$). Unfortunately, government data

on tax payments were not available to the research team.
7. This point is made clearly in Lall and Streeten (1977), p. 52.
8. Reuber *et al.* (1973) found that the sixty-four import-substituting FDI projects in their sample averaged 45 percent of their input purchases from local sources, and imported the rest.
9. Ingles and Fairchild (1978) argue that this result may be due simply to the larger average size of the foreign firms, as they discovered in Brazil, Colombia, and Mexico. In the present study, firm size was not significantly different between local and foreign firms. Lall (1978), on the other hand, reports that many studies show no difference between local firms' and MNE affiliates' profitability, despite the theoretical expectation that MNE competitive advantages would lead to higher profits.

Bibliography

Adler, M. and G. Stevens (1974) 'The trade effects of direct investment', *Journal of Finance* (May) pp. 655–76.

Agarwal, J.P. (1976) 'Factor proportions in foreign and domestic firms in Indian manufacturing', *Economic Journal.*

Areskoug, Kaj, (1976) 'Private foreign investment and capital formation in developing countries,' *Economic Development and Cultural Change*, April.

Biersteker, Thomas (1978) *Distortion or Development?* (Cambridge, Mass.: MIT Press.

Caves, Richard (1982) *Multinational Enterprise and Economic Analysis*, Cambridge: Cambridge University Press.

Courtney, W.H., and D.M. Leipziger (1975) 'Multinational corporations in less developed countries: the choice of technology,' *Oxford Bulletin of Economics and Statistics.*

Evans, Peter (1977) 'Direct investment and industrial concentration,' *Journal of Development Studies*, July.

Grosse, Robert (1983) 'The Andean foreign investment code's impact on location of foreign investment,' *Journal of International Business Studies*, Winter.

Hufbauer, Gary, and Michael Adler (1968) *U.S. Manufacturing Investment and the Balance of Payments*, Washington, DC: US Treasury Department.

Hunt, Shane (1972) 'Evaluating direct foreign investment in Latin America,' in Einaudi, Luigi (ed.) *Latin America in the 70s*, Santa Monica, Calif.: Rand Corp.

Ingles, Jerry, and Loretta Fairchild (1978) 'Evaluating the impact of foreign investment', *Latin American Research Review.*

International Labour Office (1981) *Employment Effects of Multinational Enterprises in Developing Countries*, Geneva: ILO.

Lall, Sanjaya (1978) 'Transnationals, domestic enterprises, and industrial structure in host LDCs: a survey,' *Oxford Economic Papers*, July.

———— (1980) 'Vertical inter-firm linkages in LDCs: an empirical study,' *Oxford Bulletin of Economics and Statistics*, August.

Lall, Sanjaya, and Paul Streeten (1977) *Foreign Investment, Transnationals and Developing Countries*, London: Macmillan.

Lecraw, Donald (1983) 'Performance of transnational corporations in less developed countries,' *Journal of International Business Studies*, spring/summer.

Mason, R. Hal (1973) 'Some observations on the choice of technology by multi-national firms in developing countries,' *Review of Economics and Statistics.*

May, Herbert (1974) *The Role of Foreign Investment in Latin America: Some Considerations and Definitions*, New York: Fund for Multinational Management Education.

Morley, Samuel, and Gordon Smith (1977) 'Limited search and the technology choices of multinational firms in Brazil,' *Quarterly Journal of Economics*, May.

Newfarmer, Richard, and W.F. Mueller (1975) *Multinational Corporations in Brazil and Mexico: Structural Sources of Economic and Non-economic Power*, Washington, DC: US Senate Subcommittee on Multinational Corporations.

Reuber, Grant, et al., (1973) *Private Foreign Investment in Development*, Oxford: Oxford Unviersity Press.

Stewart, J.C. (1976) 'Foreign direct investment and the emergence of a dual economy', *Economic and Social Review*, January.

Teece, David (1976) *The Multinational Corporation and the Resource Cost of International Technology Transfer*, Cambridge, Mass.: Ballinger.

United Nations (1978) *Transnational Corporations in World Development: A Re-examination*, New York: United Nations.

United Nations Centre on Transnational Corporations (1983) *Transnational Corporations in World Development: Third Survey* New York: United Nations.

United States Tariff Commission (1973) *Implications of Multinational Firms for World Trade and Investment and for US Trade and Labor*, Washington, DC: USGPO.

Vaitsos, Constantine (1974) *Intercountry Income Distribution and Transnational Enterprises*, Oxford: Clarendon Press.

Wells, Louis (1983) *Third World Multinationals*, Cambridge, Mass.: MIT Press.

Willmore, L. (1976) 'Direct foreign investment in Central American manufacturing,' *World Development.*

Chapter six

The Andean Foreign Investment Code's impact on multinational enterprises*

Introduction

Codes of conduct on multinational enterprises (MNEs) are a policy tool that gained widespread international favor during the 1970s. It was argued that, with sovereignty at bay for individual nation-states, governments needed to team up to countervail the power of transnational firms. Beginning with the Andean Foreign Investment Code in 1971, the 1970s saw ten major intergovernmental, transnational code proposals/ agreements, as well as dozens of codes enacted by companies and labor organizations. At this time it is worth reviewing the experience of intergovernmental codes, as a basis for evaluating future efforts at such regulation. Because it is both the oldest code and the only one whose rules are binding on MNEs, the Andean agreement makes an interesting focal point for analysis.

Possibly the main force behind the move toward codes of conduct came from less-developed countries (LDCs), both individually and in groups such as the Andean Pact and the Non-Aligned Countries. At the beginning of the 1970s these countries were pressing for greater involvement in decision-making at the United Nations, IMF, and other international institutions — in addition to their complaints against foreign MNEs (based in developed countries) that controlled substantial segments of their economies. These pressures led to United Nations resolutions for a New International Economic Order,[1] for a Charter of the Economic Rights and Duties of States,[2] and for creation of a Code of Conduct on Multinational Enterprises,[3] as well as to the numerous national restrictions on MNEs that the LDCs enacted. One aim of the LDCs has been and continues to be the redistribution of benefits (and costs) of company activities among the countries involved.

Because foreign direct investment (FDI) is a major vehicle used by MNEs to do business in many host countries, the issue of entry into the host country is central to any regulatory policy. From the standpoint of the host country, regulation (e.g. codes) can be used to channel investment into desired areas of industries, and to constrain the activities of foreign companies (as intended by the Andean countries). From the standpoint of the multinational investors, the provisions of any code must be reviewed to decide whether some rules are so strict as to deter investment, or other rules are so beneficial as to attract investment, in a code-adopting country. This paper will evaluate the impact of specific code policies on MNEs' decisions to locate their plants and offices in different countries. Judgment as to the desirability of the new location patterns from the standpoint of host countries is left for further analysis.

In addition to the issue of FDI location, it will be useful to consider the impact of codes on MNE operations. The benefits and costs generated by activities of MNEs can clearly be altered by changing the firms' operating policies, even without altering the country of location. The paper will thus evaluate specific code policies as they may affect MNE decisions on the firms' operations.

The Andean Foreign Investment Code (Decision 24)

The Andean Code was enacted by the Andean Pact (Bolivia, Chile, Colombia, Ecuador, and Peru)[4] on 31 December 1970 as Decision 24, 'Common Regime of Treatment of Foreign Capital and of Trademarks, Patents, Licenses, and Royalties.'[5] Decision 24, in its five chapters and fifty-five articles, sets forth a comprehensive package of rules for the entry and operation of foreign-owned businesses in the Andean Pact. Any foreign investment project must be approved by the 'competent national authority' (a body selected in each member country by the national government), which is instructed to reject investments that fail to follow Decision 24 and other Andean development guidelines.

This code contains explicit rules requiring partial fade-out of ownership after fifteen years (twenty years in Ecuador and Bolivia). Foreign investment that occurred before Decision 24 is allowed to retain foreign ownership, although the intra-ANCOM (Andean Common Market) tariff preferences are not offered to firms which do not fade out their ownerships.[6] Limits are set on profit remittance and on reinvestment of capital. Nine articles spell out the limits of foreign investors' use of patents and trademarks to protect their industrial technology. Chapter III precludes any new foreign investment in a number of nonmanufacturing industries, such as, public services, insurance, and banking, and explicitly states that each national government may restrict other sectors to 'national' investors (that is, firms at least 80 percent owned by Andean interests). Finally, the Code specifies that foreign investors must disclose several types of information to the host country's 'competent national authority,' so that accurate records can be kept by the Junta, or Secretariat, of the Pact concerning foreign investors' activities.[7]

The policies in Decision 24 are implemented through national laws in each country, and enforcement of the Code is the responsibility of each country's competent national authority.[8] Venezuela joined the Pact in 1973 and adopted the Code, while Chile withdrew in 1976 and dropped the Code.

Hypotheses

The basic hypotheses to be tested are:

Hypothesis 1: the Andean Foreign Investment Code discourages company activities within the code-adopting countries, because it raises the cost of doing business there. More specifically, the location of FDI will be biased away from code-adopting countries.
Hypothesis 2: the Code alters MNE practices in their foreign affiliates (and possibly in their home countries as well).

The first hypothesis could refer generally to any restriction on business activities that is established in potential host countries. Here it refers specifically to the fact that the Code places limitations on the multinational firms' decision-making flexibility to determine costs or revenues, that is, their ability to achieve profitability in a given country.[9]

For example, if the limit on foreign ownership of local enterprises lowers expected revenue for a potential investor, or if local borrowing constraints raise expected costs of operation, the MNE will find the code-adopting country's market less desirable than before, because either condition will lower profitability. Because the code under consideration clearly attempts through some of its provisions to alter the company-determined pattern of foreign investment and operations, it does lessen the MNE's flexibility, and thus it is expected to deter FDI entry.

One could anticipate generally that codes will tend to reduce investment coming into countries that do adopt them, and increase investment in the home country as well as in foreign countries that do not accept them. The bias goes against code-adopting countries and not against FDI in general. This hypothesis, if it can be supported, will demonstrate a serious cost in foregone direct investment (and capital accumulation) incurred by the countries adopting a code, compared to those countries that do not.

Because the Andean Foreign Investment Code was instituted in 1971, considerable post-adoption data exist for testing the first hypothesis. Although the Andean Code does not contain provisions on all of the issues that are currently under debate in various contexts, it does include many of the important issues, and its regulations are mandatory in the member countries. It was expected that this code would discourage foreign investment in ANCOM relative to that in other Latin American countries.[10] This hypothesis, if supported, will give other LDCs an important warning that use of codes may penalize them, unless all countries adopt such codes. The second hypothesis is intended to show the influence of codes on MNE behavior if the extreme result of FDI deterrence is avoided. That is, in new affiliates created after a code is passed, as well as in previously-existing affiliates, some operating policies are expected to change as a result of code stipulations. Even though a code may not deter a firm from investing in a participating country, it may affect the activities that the firm carries out in that country.

Results of testing this second hypothesis will not necessarily show a zero-sum game — that is, the countries that, through a code, reduce monopolistic business practices and gain greater access to industrial technology (among other code goals) may not cause the companies to 'lose' in terms of ultimate affiliate performance. Quite conceivably, both sides can gain from new, well-defined 'rules of the game.'

The Andean Code can be used to test this second hypothesis as well. Subsequent to its enactment, a substantial number of additional MNE subsidiaries have been established in these countries, and most previously-existing subsidiaries remain.

Method

Two major analytical techniques were used to examine the Andean Foreign Investment Code's effect on MNE decisions. First, one or two managers at each of twenty-eight MNEs (primarily US-based) were interviewed during 1978 and 1981. These managers were either senior corporate executives in the home office, responsible for Latin American or international operations, or managers of the Latin American regional offices in Miami, Florida. In each case the manager was questioned about his firm's responses to the specific provisions of the Code. Of twenty-nine companies initially selected, all but one agreed to participate (anonymously). The sample was chosen from eight industries, using the largest firms in each. The Appendix to this chapter presents some of the firms' characteristics.

The second technique used was to fit ordinary least-squares regressions to annual data related to foreign investment from 1966–81 in Latin America. Although relatively few observations are available, some useful trends can be distinguished. Additional statistical tests were performed on aggregate foreign investment data to support and extend the regression results.

Before continuing with a detailed investigation of the effects of the Code on foreign investment location, the relative priority of such regulations in foreign investment decisions should be noted. Numerous authors have arrived at the conclusion that manufacturing FDI occurs primarily to serve the host country's market. In a survey of seventy-six firms carried out by the Conference Board for the US Department of Commerce in 1972, for example, it was found that the most significant reason for choosing a country for FDI was to 'maintain or increase market share locally'; 'of most importance to the companies interviewed in making foreign investments were market factors.'[11] Thomas Horst discovered that 'in the case of a "horizontal" investment (one in which the subsidiary produces the same product for the foreign market as the parent produces for the US market) a firm's willingness to invest abroad is likely to depend on the size of the foreign market for the firm's product.'[12] Scaperlanda and Mauer, in their 1969 study of US manufacturing investment in the European Economic Community, found that among a set of hypothesized reasons for choosing an investment location, 'the size of market was the only significant variable in every case.'[13] The major importance of demand conditions in determining manufacturing FDI location also has been stressed in several other studies.[14]

The present analysis does not contradict these findings. Rather, an attempt is made to look at the incremental effects of new government regulations on foreign investment decisions. In this context the host country's market is one of the many variables to be factored out in order to focus on foreign investment codes.

Measures of the Andean Code's impact on FDI

Attempts to separate the significance of this Foreign Investment Code from other conditions and events of the early 1970s may be criticized on both empirical and statistical grounds. It is very difficult to separate the impact of the Code from the impacts of strongly nationalistic, anti-FDI governments which came to power in Latin America at the end of the 1960s — especially in Chile, Peru, and Ecuador. Similarly, Venezuela joined the Pact only in 1973, just when the OPEC countries (including Venezuela) began taking control of foreign-owned oil companies and often restricting other foreign firms through domestic policy. The oil crisis itself clearly signaled a structural shift in world economic conditions that also affected the growth and possibly the distribution of total FDI. Finally, the Code itself refers principally to manufacturing industries in ANCOM, while many of the available statistical data include all industries, and cannot easily be disaggregated.

In order to cope with these multiple and substantial difficulties, a wide variety of data sources and measures were employed. Results of several of these efforts are presented here, and others appear in Grosse (1980). In each instance, the measures of Decision 24's impact are consistent with the hypotheses under consideration.

117

US Department of Commerce data

The most detailed macroeconomic data that are available relating to FDI come from the US Department of Commerce. These data include stocks and flows of FDI and FDI-related claims world-wide, often disaggregated by country and by industry (for a few selected industries). Using data on total manufacturing for the eight Latin American countries covered (namely Argentina, Brazil, Mexico, and Panama; Colombia, Chile, Peru, and Venezuela), Grosse (1980) found weak evidence to suggest that the rate of increase of the US direct investment position in the four ANCOM countries was slower than elsewhere in Latin America.

Given the various problems with the previous measure — such as, that it includes the sum of both investment and (minus) disinvestment; that it measures book value rather than market value; and that it includes debt (local and foreign) as well as equity investment — another measure of US FDI was used. Table 6.1 presents results of a regression analysis that uses property, plant, and equipment expenditures by affiliates of US-based companies as the dependent variables.[15] This analysis uses GDP as a proxy for size of the host-country market, with the Andean Code appearing as a dummy variable. This proxy for FDI has numerous faults — especially that it includes all industries, instead of just the manufacturing industries desirable for the test — but it also gives some support to the hypothesis that the Code did deter investment, at least for a while. Note that the Code's coefficient is clearly insignificant after 1976.

The final row in Table 6.1 shows the total net inflow of FDI in all industries from all countries into the Andean Pact during 1966–75, as reported by the Inter-American Development Bank. These results are similar to those shown for US firms alone — which should not be surprising, since US firms made well over half of the total FDI in the region. (The insignificant results using these data for shorter or longer time periods are not presented.)

Harvard Multinational Enterprise Project data

Since it was founded in the early 1960s, the Harvard Business School's Multinational Enterprise Project has amassed and studied a considerable amount of information about US international business. Particularly relevant to the present analysis is the annual data concerning foreign subsidiaries of about 200 US multinational enterprises.[16] The project's databank contains information about the number of new subsidiaries established annually in various countries, from 1966 to 1975. Data are available for various industrial sectors at the three-digit SIC level. Also, manufacturing facilities are separated from sales offices, so that each pattern of investment can be examined separately.

Table 6.1 Regression results: impact of the Andean Foreign Investment Code on FDI in the region

These results come from the hypothesized relationship:
$$FDI = \beta_0 \text{ (constant term)} + \beta_1 \text{ (market size)} - \beta_2 \text{ (Andean Code)}.$$

Years	Measure of FDI	β_0	β_1	Measure of market size	β_2	Andean Code	R^2
1966–74	PPE_{US}	5.26 (0.11)	0.005 (2.40)	GDP_{t-1}	−19.02 (−1.14)	Dummy variable	0.63
1966–75	PPE_{US}	−52.91 (−2.66)	0.008 (8.54)	GDP_{t-1}	−37.07 (−3.10)	D = 0 before	0.93
1966–76	PPE_{US}	−45.28 (−3.33)	0.008 (12.44)	GDP_{t-1}	−34.34 (−3.31)	The Code	0.96
1966–77	PPE_{US}	18.35 (0.58)	0.005 (3.31)	GDP_{t-1}	−7.95 (−0.28)	D = 1 during	0.68
1966–78	PPE_{US}	22.20 (0.85)	0.004 (4.05)	GDP_{t-1}	−6.10 (−0.24)	The Code	0.75
1966–79	PPE_{US}	2.85 (0.11)	0.005 (5.38)	GDP_{t-1}	−16.66 (−0.62)		0.81
1966–80	PPE_{US}	21.00 (0.93)	0.004 (5.25)	GDP_{t-1}	−3.32 (−0.12)		0.80
1966–75	FDI	−813.15 (−2.14)	0.001 (2.65)	GDP_{t-1}	−452.19 (−1.61)		0.50

Sources: Computed from US Department of Commerce, *Survey of Current Business*, various issues; Inter-American Development Bank, *Economic Progress in Latin America*, various years and IMF, *International Financial Statistics*, various issues.
(t-statistics in parentheses)
Notes:

PPE_{US} =	Property, plant, and equipment expenditures by affiliates of US-based corporations in Andean Pact countries (in current dollars).
GDP_{t-1} =	Gross domestic product of Andean Pact countries for the previous year (in current dollars).
FDI =	Net inflow of foreign direct investment from all countries into the Andean Pact (in current dollars).

The Code's impact was strongest in the first three years shown; after that the coefficient's t-statistic is very insignificant, that is, the Code's impact is not significantly different from zero. These results are consistent with those in Grosse (1980), a study which used three different measures of FDI.

Statistical investigations using these data must be somewhat limited. No measure of the value of each investment project is available — there is only the number of subsidiaries that were formed. As with the Department of Commerce data, there are only a few years (observations) available for analysis in each time series. Despite these drawbacks a few elementary tests can be performed. In Table 6.2 the results of x^2 tests between the number of subsidiaries entering each groups of countries, before and during the existence of the Andean Code are presented.

As expected, the x^2 tests support the hypothesis that Decision 24 (the Code) has reduced the rate of new foreign investment in the Andean Pact for the particular time period available. Even with recalculation of the statistic to eliminate bias caused by inclusion of Chile and Argentina, where extreme leftist governments took power in 1970 and virtually stopped all incoming FDI (as well as nationalizing many existing investors), and with separate tests of adjusted time periods, the x^2-tests yield highly significant results. In each of the tests, the ANCOM countries were found to have attracted significantly fewer new subsidiaries during the early existence of the Code, relative to the non-ANCOM Latin American countries. Unfortunately, the dataset does not contain information after 1975, so it is not possible to measure the expected drop off in the code's effect that was seen in previous data from other sources.

Table 6.2 Number of subsidiaries established by US, MNEs, before and during the Andean Code

Country	1966–70	1971–5			
Bolivia	2	3	$x^2 = 9.95$		
Chile	28	8		1966–70	1971–75
Colombia	31	27			
Ecuador	18	12			
Peru	32	9	ANCOM	195	122
Venezuela	81	63			
ANCOM	195	122	Other		
			Latin American	482	391
Argentina	82	17			
Brazil	120	145	x^2critical $x^2 1,.05 = 3.84$		
Costa Rica	23	14			
Guatemala	21	20	Eliminating Argentina and Chile		
Mexico	179	149			
Nicaragua	9	9	$x^2 = 4.982$		
Panama	39	3	x^2critical $= x^2 1,.05 = 3.84$		
Uruguay	9	4			
non-ANCOM Latin America	482	391			

Source: Computed from J. Curhan, W. Davidson, and R. Suri, *Tracing the Multinationals* (Cambridge, Mass.: Ballinger, 1977), as shown in Grosse (1980), p. 104.
Note:
The x^2 tests compare the total investment in ANCOM, pre-1971 and post-1970, with the total investment in other Latin American countries in each time period. A test value greater than the critical value calls for rejecting a hypothesis that the two groups of countries had equal changes in investment between time periods (at the 95 percent confidence level).

The data used in the two samples above have a variety of drawbacks, as noted. For purposes of understanding the reactions of foreign investors to codes of conduct on multinational enterprises, some additional information is desirable. Views of managers in the firms themselves offer a third and final source of empirical evidence. Interview data also will allow for examination of specific code provisions as they affect or do not affect the firm's decisions.

Interview data

Company interviews held during 1978 and 1981 contained a variety of questions about the contents of Decision 24 and other Andean Pact regulations on FDI. From these discussions, it was found that the list of eight major issues presented in Table 6.3 does contain the areas of most friction in the government/company negotiations on foreign investment entry.

Table 6.3 Interview results: significance of Code provisions in foreign investment location decisions

	Important	Some effect	No effect
1. Financial restrictions (such as limits on profit remittance and reinvestment)	95	5	0
2. Acquisitions	31	31	38
3. Ownership	29	29	41
4. Technology transfer	*	*	*
5. Information disclosure	0	16	84
6. Antitrust	0	5	95
7. Transfer pricing	0	5	95
8. Labor relations	0	0	100

Source: Interviews at twenty-eight MNEs in 1978.
* No significance was attached to any specific provisions regarding technology transfer in the Andean Code. However, several managers consider that the uncertainty about industrial property protection in the Andean Pact is a drawback to new investment there.

In the Code itself, however, not all eight of the issues appear. For example, there is no specific mention of corrupt business practices or employment policies, nor is there much regulation on competition. (The basic antitrust rules for the Andean Pact are in Decision 45.) According to the MNE managers who were interviewed, three provisions of the Code present serious problems to the US investor:

1. Ownership requirements — new investment must be scheduled with a fade-out agreement to at most 49 percent ownership within fifteen years (twenty years in Bolivia and Ecuador).
2. Limit for existing majority-owned investors of 7 percent unrestricted annual addition to registered capital — thus, firms which invested before Decision 24 are being pushed to enter the fade-out scheme just like new investors. (Reinvestment of profits

earned by foreign enterprises is considered to be 'new investment', and is covered by the rules in 1 above.)

3. Profit remittance limit — majority-owned investors may remit up to 20 percent of their registered capital i.e. 20 percent of their percentage of total registered capital[17] each year.

Ownership requirements

The most important rule according to the MNE managers is Article 30, which sets forth the plan for eventual divestment of at least 51 percent ownership to local investors. Articles 38–44 identify the industries in which no new foreign investors will be allowed (such as, public services, electricity, telecommunications, insurance, banking, and the media).

Table 6.4 Ownership at entry of US manufacturing subsidiaries, before and during the Andean Code

Country	1967–9 Wholly-owned	JV	1973–5 Wholly-owned	JV
ANCOM	85	48	46	29
Bolivia	2	1	1	2
Chile	11	9	4	4
Colombia	17	4	7	7
Ecuador	9	2	3	5
Peru	16	9	3	1
Venezuela	30	23	28	10
Latin America (except ANCOM)	270	105	172	98
Argentina	47	8	4	1
Brazil	52	27	51	45
Costa Rica	11	5	12	1
Guatemala	9	3	13	1
Mexico	63	43	39	29
Nicaragua	2	1	5	2
Panama	24	2	22	2
Uruguay	4	1	3	0

Source: Computed from J. Curham, W. Davidson, and R. Suri, *Tracing the Multinationals*, taken from Grosse (1980)

Note: The x^2-tests compare the proportion of joint ventures (JVs) in total investment within the set of countries before and after the initiation of the code. A test value greater than the critical value rejects a hypothesis that the proportion of joint ventures (as opposed to wholly-owned subsidiaries) has not changed since the code's inception. A 95 percent confidence level is used. The test of the full hypothesis,

$$Ho: \frac{\text{wholly-owned subsidiaries}}{\text{all subsidiaries established}} \; 1967\text{–}9 = \frac{\text{wholly-owned subsidiaries}}{\text{all subsidiaries established}} \; 1973\text{–}5.$$

yielded $x^2 = 28.56$ for ANCOM countries. The critical value in this instance is $x^2 = 11.07$, so the null hypothesis is rejected. This conclusion supports our reasoning that the use of joint ventures for new FDI increased in ANCOM since adoption of the code. A comparable test for non-ANCOM Latin America yielded $x^2 = 0.61$, with a critical value of $x^2 = 14.07$. This test shows that no significant shift toward the use of joint ventures has occurred in the rest of Latin America during the period examined (for U.S. MNEs).

Articles 12 and 13 put pressure on existing foreign investors in ANCOM to restructure subsidiaries along the same lines as new investors are doing. Investors already in ANCOM by 1 July 1971, however, are not required to divest any ownership unless they want to use the customs union tariff benefits, or unless they are in restricted industries (in which case at least 80 percent ownership must be sold to national investors). Ownership requirements such as this fade-out scheme present a serious concern to the MNEs. Formerly able to direct international operations from a central (e.g. US) location, they are now being pulled into positions of minority partnership in their foreign ventures. The full ramifications of such forcibly decentralized ownership and management are currently developing and are beyond the scope of this study.

One aspect of ownership that can be measured with available data is the use of joint ventures in new FDI. Data from the Harvard MNE Project lend some support to the presumption that Decision 24 has pushed US investors into the use of joint ventures, away from wholly owned subsidiaries. Table 6.4 shows results of two x^2-tests comparing foreign investment entry in wholly owned subsidiaries versus joint ventures during 1967–79 and 1973–75. This pair of tests demonstrates that the composition of ownership at entry of subsidiaries has shifted significantly within ANCOM from wholly owned subsidiaries toward joint ventures; in the other group of Latin American countries, no significant shift occurred.

Financial rules

Article 13 (as amended) allows unrestricted reinvestment of profits by 'foreign enterprises' up to 7 percent of registered capital. This means that if profits are greater than 7 percent of the company's capital as registered with local authorities, there is excess money available which can only be used in registered investment if the firm is willing to begin divesting at least 51 percent (unless specifically authorized by national authorities.[18] Fortunately, Article 37 (as amended) allowed for profits up to 20 percent of registered capital to be remitted annually. Thus another outlet exists for the subsidiary's income. Once both of these limits have been reached, however, the firm cannot use further profits except in local bank deposits, local portfolio investment, and unregistered reinvestment. This may mean that further profits effectively cannot be used to any advantage; bank deposits which pay as much as 50 percent per year (in Peru) are unattractive, given annual inflation rates of 80 percent and higher (also in Peru); unregistered capital cannot be repatriated even when the subsidiary is sold. Of all of the financial constraints imposed in Decision 24, the profit remittance and reinvestment limits were reported as the key problems to MNE managers.

123

As a result of Articles 13 and 37 together, existing foreign investors are strongly encouraged to follow the plan of fade-out in ownership which is applied to new investors. Once a company reaches 49 percent ownership (i.e. 51 percent local ownership), the financial limits no longer apply. These constraints were reported by the MNE managers to be serious deterrents to direct investment, as discussed and measured above. Additionally, the managers unanimously agreed that the financial rules would alter some of the firms' operating policies. Namely, profit remittance would be reduced; loan interest payments would be reduced (under Article 16 of Decision 24); and payment of royalties and fees would be cut (under Article 21 of Decision 24). Managers' assertions about their operating policies were substantiated somewhat by US Department of Commerce data on affiliate-to-parent financial flows.

A measure of annual payments of dividends and interest (plus retained earnings of unincorporated affiliates) by US affiliates in eight Latin American countries has been compiled since 1966. Likewise, data on fees and royalty payments to parent firms have been collected. Table 6.5 presents some of these data aggregated to annual average flows, for three years just before Decision 24 and for three subsequent years when both Venezuela and Chile were members of the Pact.

Table 6.5 Financial flows from Latin American affiliates to US parent firms (average annual flows in US$m)

	Income received by parent*		Fees and royalties received by parent		
	1967–9	1973–5	1967–9	1973–5	
Argentina	93	53	25	26	Mann-Whitney U-tests
Brazil	58	109	23	30	for $n_1 = n_2 = 4$.
Mexico	58	122	59	96	$U_{income} = 5 \rightarrow$
Panama	58	118	12	40	probability = 0.243
Chile	129	5	20	4	$U_{fees} = 1 \rightarrow$
Colombia	15	53	16	10	probability = 0.029
Peru	97	6	15	15	
Venezuela	408	453	28	27	

Source: US Department of Commerce, 'Selected data on US direct investment abroad, 1966–76' (xeroxed computer printout, reprinted in *Survey of Current Business*, various issues.
Note:
The Mann-Whitney U-tests compare percentage changes in income (and royalties and fees) before and during the Code between the four Andean and the four other countries. The test statistic represents the probability that the ANCOM financial flows are not stochastically smaller than the other countries' flows.
* Income includes dividends and interest payments from all affiliates plus reinvested earnings of unincorporated affiliates.

Note that the Mann-Whitney U-test on fees and royalties, between ANCOM and other Latin American countries, shows a high probability that such payments from ANCOM affiliates declined relatively during the early years of the Code. (These payments did not drop to zero, however, because some fees were not subject to Decision 24, and exceptions to the rules were granted to many foreign affiliates that pre-dated the Code.) The test on income receipts by parent firms is insignificant, though Chile and Peru did show huge declines in such flows in the latter period. Many rationales are possible for this result, but the most obvious is that Colombia's government simply allowed much greater payments of interest and dividends in the 1973–5 period. Overall, the evidence presented here on financial flows and ownership tends to support the hypothesis that company operating policies have been affected by the Code.

Implementation

One major factor that does not appear in any of the preceding discussion is implementation of code rules. Because the Andean Foreign Investment Code has been enacted through a separate (and un-equivalent) national law in each member country, its effects cannot be uniform in the region. Careful study of the enacting legislation shows a wide variance just in the reformulation of Decision 24's provisions by each country. Interview results showed that enforcement of the national laws themselves is unequal across member countries (Peru's enforcement has been and remains much stricter than Ecuador's, for example).

A few other studies have appeared and shed more light on the subject of implementation. Richard Robinson discovered, through his interviews with government officials and company managers in each ANCOM country in 1974, that the implementation of Decision 24 varied widely from country to country.[19] He found that the laws enacted by each member country have failed to reproduce the provisions in Decision 24 — in most cases allowing more lenient controls. Robinson ranked Peru as the most restrictive country, followed by Venezuela; and Chile (until it withdrew from ANCOM) as the most open, preceded by Colombia. His detailed analysis of the national legislation provides useful additional insights into the Andean Pact and Decision 24.

Altogether, the impact of the Andean Code appears initially to have been substantial both on entry decisions and on operating flexibility of MNEs. The amount of incoming FDI evidently slowed in the first few years of the code's existence, and a significant shift in ownership at entry has occurred in favor of joint ventures (with host-country partners) and away from 100 percent foreign MNE ownership. In addition, foreign investors are limited tremendously in their ability to move funds out of

ANCOM host countries. Profit remittance is limited; loans from the parent may charge no more than 3 percent above the prevailing rate in the country whose currency is used (Article 16); and no payment of royalties is allowed to the parent firm for technology transfer (Article 21). Also, entry through acquisition of existing nationally owned firms is disallowed (Article 3). These examples mark the major, but not the only, restrictions that have constrained foreign investment activities in the Andean Pact since 1971. (It again should be noted that not all of these restrictions are being implemented, and some are applied differently, in the five ANCOM countries.)

Conclusions

Evidence compiled in the above analysis gives some support to the hypothesis that codes on MNE's can alter company behavior in directions desired by the code-adopting governments. Certainly in the case of the Andean Pact, the mandatory rules imposed on foreign investors have led to changes in their ownership and financial policies.

The evidence concerning codes' effects on FDI location is less clear. Company interview responses (related to the Andean Code) lead one to believe that strict limits on ownership and financial policies deter foreign investors from code-adopting countries. The various statistical measures employed here tend to show a significant reduction of FDI into the ANCOM countries during the first five years of the Code's existence — but no enduring impact beyond that.

Regardless of the actual impact of a code on foreign investors, the code-adopting countries must be aware that the perceived impact may be equally important. The costs of a perceived reduction in FDI by Chile were sufficient to lead that country to withdraw from the Andean Pact — and to enact attractive legislaton for foreign investors as well. (This example is probably overstated, however, because Chile also was striving to overcome the drastic drop in FDI due to the 1970–3 Allende regime.) On the other hand, if code-adopting countries demonstrate a clear willingness to make exceptions to the controls, the perceived impact of a code may be small and thus insignificant to foreign investors.

As we move into the 1980s, it seems likely that constraints on foreign ownership of domestic enterprises (especially in LDCs) will follow Andean-type models rather than return to greater freedom for MNEs. Mexico has followed ANCOM in requiring local ownership; other LDCs in Africa, Asia, and Latin America are leaning in this direction by offering incentives to firms which take local partners; even developed nations such as Canada are reviewing applications for FDI with a bias toward those with local participation. Similarly, limits on the practice of acquiring existing firms, rather than establishing new enterprises, are spreading through the LDCs and into the developed countries, too. In all, the constraints facing MNEs in their attempts to undertake foreign investment include ownership and acquisition limits in a wide range of countries — code-adopters thus should not suffer less FDI relative to other nations just because of these rules.

The same trend exists in the area of financial restrictions. Serious balance of payments difficulties, especially as a result of OPEC oil pricing, are causing developed and less-developed nations alike to use capital controls aimed at limiting MNE international financial transfers. If these restrictions appear widely in national legislation, then code-adopting countries can implement similar controls without damaging their relative attractiveness to investors. Viewed against a backdrop such as we see in the early 1980s, the Andean Code appears more as an example of prevailing regulatory conditions than an extreme model of LDC nationalism.

Appendix to Chapter 6

Characteristics of the twenty-eight firms in the sample

Industry	SIC number	Number of firms	Average annual sales, 1981 (US$m)	Average number of South American nations where firm is incorporated, 1980
Chemicals	281, 287	7	9,624[a]	6
Cosmetics and health-care products	284	3	4,719	6
Electrical appliances	363–5	2	15,037	7
Food processing	204, 207	6	5,278[b]	4
Office and computing machinery	357	2	16,238	8
Pharmaceuticals	283	3	2,568	6
Telecommunications equipment	366	2	14,166	7
Tires	301	3	4,866	9

Sources: Company annual reports, interviews.
Notes:
[a] Does not include sales of one firm which is a wholly-owned subisidiary of an oil firm.
[b] Does not include sales of one privately owned firm.

Postscript to Chapter 6

In May of 1987 the Commission of the Andean Pact formally terminated Decision 24 and several subsequent rules on participation of foreign investors in the five member countries. These rules were replaced by Decision 220, which calls for essentially equal treatment of foreign and domestic capital by Andean governments, explicitly to try to stimulate foreign investment to help generate income and employment in the region.

This dramatic shift in regulation of foreign direct investment can be seen as the result of a major shift in the bargaining strengths of the governments and the foreign companies during the Latin American debt crisis of the 1980s. Specifically, as the Andean countries fell further and further into crises of servicing their foreign debt during the 1980s, they increasingly opened their economies to participation by foreign firms. Reduced limits on foreign direct investment, creation of debt–equity swap programs to redeem some of the foreign debt, and efforts to promote more exports of non-traditional products all have resulted from the crisis situation.

This outcome is clearly consistent with the bargaining theory of government–business relations (and particularly with Hypothesis 9 from Chapter 4). Historically, each time the region has fallen into a period of declining growth (i.e. recession or depression), the regulatory structure has shifted to much greater openness toward foreign firms. The consequence of this line of reasoning is that, once Latin American economies escape the overhanging debt crisis, and re-establish sustained growth paths, they will be likely to once more impose stricter rules on participation by foreign direct investors.

Notes

Acknowledgements: The author wishes to thank Jack N. Behrman, Eulogio Romero, and an anonymous referee for comments on earlier drafts. This research was supported by a Fulbright grant and a grant from the Fund for Multinational Management Education.

*This article is reprinted with permission from the *Journal of International Business Studies* (Winter 1983).

1. United Nations Resolution 3201, May 1974.
2. United Nations Resolution 3281 (XXIX), December 1974.
3. United Nations Resolution 1721 (LIII) of July 1972, called for formation of a 'Group of Eminent Persons' to decide how the UN could make recommendations regarding MNEs.
4. Venezuela participated in negotiations to create the Andean Pact, but failed to accept the founding treaty (Acuerdo de Cartagena) in 1969. On 13 February 1973, Venezuela officially joined the group. The government of Chile decided that the policies of the Andean Pact were inconsistent with that country's development plans, and on 5 October 1976, Chile withdrew from the group.
5. For another analysis of Andean Pact rules on MNEs and of the Code in particular, see Robinson [1976].
6. In the Cartagena Agreement of May 1969, which formed the Andean Pact, Chapter 5 states the rules for liberalizing intra-ANCOM trade. With several exceptions, including trade under any sectoral programs, tariffs are being

reduced at 10 percent per year from their original levels, beginning 31 January 1970.

7. Since the Code was enacted, Decisions 37 and 37a of June and July 1971 have modified it slightly. Decision 103 of October 1976 raised limits on profit remittance from 14 to 20 percent of registered capital and on reinvestment in registered capital from 5 to 7 percent.

8. Stanley Rose examined the implementation of the code in each country specifically as it concerns profit remittance, capital reinvestment, fadeout of ownership, and payment for technology transfers. His notes on specific implementation of the rules generally support the present view. See Rose, Stanley, 'The Andean Pact and its Foreign Investment Code: need for clarity?' *Tax Management International Journal*, January 1975.

9. The issue of profitability is not explored here. Because of the vast possibilities for moving profits among affiliates in MNEs, measures of accounting profits in given countries would not necessarily reflect real contributions to the total firm. Thus, while the Andean Code is expected to constrain firms' profitability, the direct results are not measured here. Instead, this analysis focuses on entry decisions and operating policies.

10. Mexico unilaterally implemented a foreign investment code under its law of 9 March 1973. This set of rules is less restrictive, however, than the ANCOM code. Extensive clarification of acceptable activities appeared in General Ruling No. 16 (6 September 1977) of the National Commission on Foreign Investment.

11. U.S. Department of Commerce, *The Multinational Corporation: Studies on US Foreign Investment* Washington US Government Printing Office, vol. 2, March 1972, p. 6–7.

12. T. Horst, 'Firm and industry determinants of the decision to invest abroad: an empirical study,' *Review of Economics and Statistics*, August 1972, p. 259. Horst's data included 1,191 US manufacturing corporations.

13. A. Scaperlanda, and L. Mauer, 'The determinants of US direct investment in the EEC.' *American Economic Review*, September 1969, p. 563.

14. See, for instance, R. Stobaugh, 'Where in the world should we put that plant?,' *Harvard Business Review*, January–February 1969, p. 130; H. Schollhammer, 'Locational strategies of multinational firms,' *Study No. 1*, Center for International Business, Pepperdine University, California, 1974, pp. 13–14, 19, US Tariff Commisson, 'Implications of multinational firms for world trade and investment and for US trade and labor', report to the Senate Committee on Finance, USGPO, February 1973, pp. 20, 102, 115.

15. This measure of FDI was first used, to my knowledge, by Stevens (1969).

16. For details on the firms' characteristics and their names, see J. Curhan, W. Davidson, and R. Suri, *Tracing the Multinationals*, Cambridge, Mass.: Ballinger, 1977.

17. 'Registered capital' basically includes that value of plant and equipment originally invested plus any incremental investment that the national authority accepts for registration. Decision 103 of the ANCOM Commission raised the limit on profit remittance from 14 to 20 percent of registered capital

(in October 1976) and it raised the limit on additions to registered capital from 5 to 7 percent.
18. The national authorities in Peru and Colombia thus far have been unwilling to allow reinvestment in registered capital above the 7 percent limit, without forced fade-out of ownership.
19. Robinson 1976, Ch. 8.

Bibliography

Behrman, Jack, 'Sharing international production and international industrial policy', *Law and Policy in International Business*.

Business International (1971) *Investing Licensing and Trading Conditions Abroad*, New York: Business International Corp.

Chance, Steven (1978) 'Codes of conduct for multinational corporations', *The Business Lawyer*, April.

Diaz, R. (1971) 'The Andean Common Market: challenge to foreign investors,' *Columbia Journal of World Business*, July–August, pp. 22–8.

Franko, Larry, (1974) 'International joint ventures in developing countries: mystique and reality,' *Law and Policy in International Business*, pp. 315–36.

Furnish, Dale, (1972) 'The Andean Common Market's regime for foreign investors,' *Vanderbilt Journal of Transnational Law*, spring.

Grosse, Robert (1980) *Foreign Investment Codes and Location of Direct Investment*, New York: Praeger.

Meeker, Guy 'Fade-out joint venture: can it work for Latin America?' *Inter-American Economic Affairs*, spring.

Middlebrook, Kevin (1978) 'Regional organizations and Andean economic integration, 1969–75,' *Journal of Common Market Studies*, September, p. 62–82.

Robinson, Richard (1976) *National Control of Foreign Business Entry*, New York: Praeger.

Rubin, Seymour (1971) 'Multinational enterprises and national sovereignty,' *Law and Policy in International Business*.

Schwamm, Henri, and Germidis, Dimitri (1977) 'Codes of conduct for multinational companies: issues and positions,' European Centre for Study and Information on Multinational Corporations, Brussels.

Stevens, Guy (1969) 'Fixed investment expenditures of foreign manufacturing affiliates of US firms: theoretical models and empirical evidence,' *Yale Economic Essays*, spring, pp. 137–200.

United Nations (1978) *Transnational Corporations in World Development: A Re-Examination*, New York: United Nations.

Vagts, Detlev (1973) 'The host country faces the multinational enterprise,' *Boston University Law Review*, March.

Vaitsos, Constantine (1973) 'Foreign investment policies and economic development in Latin America,' *Journal of World Trade Law*, November–December.

Vernon, Raymond (1977) *Storm Over the Multinationals*, Cambridge, Mass.: Harvard University Press.

Waldmann, Raymond (1980) *Regulating International Business Through Codes of Conduct*, Washington, DC: American Enterprise Institute.

Wallace, Don, (ed) (1974) *International Control of Investment*, New York: Praeger.

Chapter seven

Resolving Latin America's transfer problem

Introduction

The crisis of external indebtedness in Latin America has disrupted much of international business in the region for the past six years. As public- and private-sector borrowers in Latin America have encountered persistent difficulties in meeting foreign-debt obligations, it has become more and more clear that some substantial shift in the manner of dealing with the crisis is needed. This chapter reviews the major issue involved in the crisis, namely the 'transfer problem' that requires a shift of real resources from the debtor countries to their foreign creditors. An attempt to estimate future ability of the major debtor countries to repay their foreign debts shows the unlikelihood of meeting the existing obligations under current conditions. The failure of banks and governments to deal adequately with this crisis sets the stage for new major problems in the near future.

The Latin American debt crisis possesses many of the characteristics of previous major international financial crises in this century. In each case, the crisis or problem can be viewed as a 'transfer problem,' necessitating some movement of real resources from debtor to creditor countries in the international financial system.[1] In the current case, most Latin American countries have incurred levels of indebtedness which require exceedingly high repayment costs relative to the countries' abilities to repay. The transfer problem refers to their need to generate outflows of real purchasing power in the form of foreign currency or resources such as exports and/or ownership of domestic assets (i.e. foreign investment) sufficient to equal debt-servicing charges.

Several other major transfer problems in international finance occurred during the twentieth century. In the 1920s, Germany faced a demand by the Allied countries to pay reparations for damages caused during the First World War. This charge against Germany called for payment of annual reparations in foreign currency, which became a transfer problem as Germany sought to pay for more than a decade. The foreign debt of Latin American countries in the 1920s and 1930s can be considered in this same category, since the timing and difficulties of repayment were very similar. Next, the OPEC oil crisis of the 1973-4 period,[2] which continued through the rest of the 1970s, created a massive transfer of purchasing power to the oil-exporting countries, and necessitated in return a transfer of real wealth by the importing countries. The OPEC crisis itself is a fundamental base of the final transfer problem to be discussed — the current Latin American debt crisis.

The chapter begins with a discussion of the transfer problem. Some of the key characteristics of the previous three transfer problems in this century will then be sketched. Following this, focusing on the current Latin American problem, projections are made of several countries' abilities to repay debt over the next five years. Both the history and the projections lead to a conclusion that some of the existing foreign debt will be unpaid, through default, loan devaluation, government policy intervention, and/or inflating away some of the real value. The basic result of the analysis is a conclusion that the problem will linger indefinitely — but some promising steps have been taken by the private sector that could well eliminate the crisis nature of the problem and lead to renewed commercial bank lending and foreign investment into the region.

The transfer problem

Definition

While the need for large-scale international financial transfers has occurred before this century, the term *transfer problem* was coined only at the time of the German reparations problem. The Dawes Commission (which established reparations payments and scheduling), and later, John Maynard Keynes used the term to refer to part of the reparations issue:

> The Dawes Commission divided the problem of German Reparations into two parts — into the *Budgetary* problem of extracting the necessary sums of money out of the pockets of the German people and paying them to the account of the Agent General, and the *Transfer* Problem of converting the German money so received into foreign currency.[3]

A second aspect of the transfer problem is the *timing* of the transfer. In the German reparations case, even the most optimistic observers expected that part of the requisite transfer of purchasing power would be postponed by international borrowing done by Germany. This borrowing allowed payment as required under the terms of the reparations agreements, but created new indebtedness which would have to be repaid in the future. Such a strategy in the extreme could allow the transfer to be 'financed' indefinitely, presumably until sufficient output were created in the future to generate foreign exchange for ending the debt. As in the case of current Latin American debt, new borrowing could be used to finance old interest and principal repayments until some creditor(s) refused to continue the game and caused a crisis, or until the debtor country's 'belt-tightening' allowed the transfer to be completed.

Solving the transfer problem

Looking mainly at the problem of generating sufficient exports to pay the transfer, Keynes (1929) originated the *terms of trade* or *elasticities approach*. This view focused on the point that Germany (to pay reparations) or any other debtor country would need to increase its exports dramatically, such that foreign exchange could be earned to make payment on the contracted debt/reparations. As exports increased, there would be a downward bias on export prices, since only such a price reduction would attract additional buyers of these products. Overall, it was not clear that increased exports really would generate sufficient foreign exchange, because the result depended completely on the price elasticity of demand for exports. If export demand was price inelastic, then dropping prices (either through direct price cuts or through currency devaluation) would only have increased the quantity of exports, without raising the value of export sales. This 'elasticity pessimism' led Keynes and others to doubt that Germany could ever make reparations payments as required.

Looking at income effects of spending by debtor and creditor nations, Ohlin (1929) stated the *modern view* or *absorption approach*. In this view, adjustments in both lending and borrowing countries' real incomes need to be considered to judge the ultimate effect of the transfer effort. When a debtor country transfers some of its purchasing power to a creditor, national income will decline in the debtor country and rise in the creditor country. This shift in income will make the creditor's demand for debtor-country exports rise, as long as income elasticity of demand for those exports is positive. Similarly, the debtor country's residents will demand less imports along with the decline in their purchases of domestic goods (as GDP falls), so the debtor's trade balance will improve. The income effect tends to help carry out the transfer, while the price effect tends to hinder it. Whether or not the transfer can be made depends on both effects. The feasibility of the transfer depends completely on the empirical situation involved.

More recent studies of the transfer problem generally have focused on evaluating transfer possibilities in various empirical situations (e.g. Balogh and Graham 1979, who investigated the OPEC crisis), and on some of the theoretical specifications of the problem itself (e.g. Johnson 1976).

The balance-of-payments view

Before moving on to empirical examples, let us look at one more view of the transfer problem. This view attempts to specify the possible outcomes of a transfer problem. Simply, of course, the possibilities are that the transfer either will or will not be made — but what time frame should be allowed for the full transfer to occur? And what means of payment will be accepted as accomplishing the transfer? The balance-of-payments view illuminates these issues.

The balance-of-payments view shows what steps are available to equilibrate a country's international transactions, including the transfer. Equilibrating the balance of payments means making the transactions balance; it does not necessarily mean stabilizing the country's payments in any one category of transactions. Brazil's balance of payments for 1986, as shown in Table 7.1, offers a useful example for examining the issues.

Table 7.1 Brazil's balance of payments for 1986 (US$m)

	Inflows	*Outflows*	
Imports, merchanise	12,866	22,393	Exports, merchanise
Trade in services, net	12,463		
interest payments	9,093		
other services	3,370		
Unilateral transfers, net		87	
Direct investment, net	115		
Long-term loans		13,221	
Short-term capital flow	1,334		
Errors and omissions		210	
Official reserve flow		3,330	

Source: Banco Central do Brasil.

Debt-servicing needs are shown as interest payments ($9 billion) plus an unrecorded amount of expense for fees involved with renegotiations of loan contracts (since principal payments are *not* being made when due). This means that Brazil faced an external debt-servicing requirement (i.e. a transfer problem) of about $10 billion per year, even without

135

repaying any principal on foreign debt. Beginning from this fact, the balance-of-payments view points out some of the options available to pay for or finance the debt.

First, as is widely discussed in the literature, Brazil can run a merchandise trade surplus, generating foreign exchange to pay the transfer. In fact, in 1986 the trade surplus was about $9.5 billion, which covers about 95 percent of the amount needed. Second, Brazil can try to 'finance' the problem by borrowing new foreign exchange to meet old obligations. While this strategy postpones the problem, rather than reducing it, Brazil did borrow about $9 billion in 1986. (Only about $3 billion was new financing in 1986; the rest was refinancing of existing foreign loans.) Together with the trade surplus, this borrowing adequately covered the transfer needs for the year. Notice, however, that several more strategies may be followed here. Moving down the balance-of-payments table, Brazil could try to increase exports of services; and/or Brazil could seek additional unilateral transfers (gifts) from abroad; more foreign direct investment could be attracted (there was net *disinvestment* in 1986); and perhaps more official reserves could be created by the IMF. Any of these choices wold help carry out the transfer, with the *caveat* that direct investment will create future service imports as some profits are remitted to the home country of the investor. On the other hand, direct investment will create more national income, which may generate more products for export. The overall long-term impact of additional direct investment on the balance-of-payments remains ambiguous.

The balance-of-payments view moves our focus away from evaluation of price and income effects toward choice of methods to deal with the transfer problem. Notice that one method of dealing with the problem has not been mentioned so far. In principle it is possible for a debtor country to renounce responsibility for making the transfer. This possibility, that is, default, has been utilized in the past, and remains a threat that the debtor countries can bring to bear when pressured by foreign creditors. In fact, the costs to the debtor countries in terms of lost business and economic sanctions make complete default a highly unlikely strategy in the 1980s.

In sum, the transfer problem is viewed here as a difficulty in transferring acceptable financial or real resources from debtor to creditor countries when such a BOP imbalance exists. Since national balances of payments (on essentially every basis, i.e. trade, current account, basic, official reserve transactions, etc.) are generally in deficit or surplus, the transfer problem in principle exists continually. The focus of analysis here is on the most recent crisis which constitutes a severe transfer problem during the 1980s. The transfer itself explicitly is not limited to financial instruments, but also may involve sales of real resources, for example through foreign direct investment or even through barter trade.

Three empirical transfer problems

German reparations

In the aftermath of the First World War, the European Allied countries determined to exact from Germany a large sum of money to compensate partially for damages done in the war. Initially, the size of these reparations payments was not fixed — the demands were essentially open-ended. In 1921 the Reparations Commission established a figure of 132 billion gold marks. In 1924 the Dawes Commission set a plan calling for annual payments of 1 billion gold marks for five years and 5 billion gold marks annually thereafter. The Young Commission in 1930 revised both the size of the annual transfers and the number of years to termination of the charges. Table 7.2 shows the full set of reparations payments made from 1918 to 1931. Notice that the German reparations were a transfer problem arising not out of an original loan, but forced upon Germany as a unilateral transfer.

Table 7.2 German reparations payments (in billions of gold marks*)

Payments made	Estimates made by Reparations Commission	Estimates made by German government
11 November 1918 to 31 August 1921	9.7	42.1
Under the Dawes Plan (Sept. 1924 to Mar. 1930)	7.6	8.0
Under the Young Plan (Apr. 1930 to June 1931)	2.8	3.1
Other	0.8	14.6
Total	20.9	67.8

Source: Kindleberger 1973, p. 35.
*The Reichsmark was worth approximately US$0.238 during the Dawes and Young Plan periods.

The results of the German reparations were that a substantial transfer of real resources took place from Germany to the Allied countries during the years after the First World War. Also, a huge volume of lending took place from the Allied countries to Germany, thus allowing Germany to avoid carrying out the full transfer at that time. The German

137

hyperinflation of 1921–2 did not reduce the burden of reparations, since their value was set in gold marks, that is effectively in gold itself. When the global depression hit in 1929, Germany faced even greater difficulties in making the transfer, since demand for exports fell and international lending declined. Ultimately, in 1931, Germany unilaterally renounced the obligation to repay, and reparations terminated.

Latin American debts in the 1920s and 1930s

Compare the German experience with those of the Latin American countries in the 1930s. During the 1920s, in Latin America as well as in Western Europe, the United States loaned substantial funds to public- and private-sector borrowers. In both cases, most (over 95 percent) of the lending was to government borrowers, and most was done through foreign bond issues in New York. Table 7.3 shows the extent of lending to various Latin American borrowers through bond issues in the United States. Notice that one major Latin American country, Mexico, does not appear as a borrower in the 1920–31 period. This is because Mexico remained in default on bonds contracted during 1900–14, and no new lending of this type was undertaken after the war.[4]

Table 7.3 Distribution of Latin American bond issuers in the United States, 1920–31 (US$m)

Country	National government	State or local govt. or govt guaranteed	Private sector	Total
Argentina	552.4	188.8	0	742.2
Bolivia	38.3	0	0	0
Brazil	159.4	230.5	4.0	393.9
Chile	191.5	109.5	0	301.1
Colombia	56.1	114.7	23.0	193.8
Cuba	110.6	0	0	178.2
Panama	20.1	4.0	0	24.1
Peru	91.0	4.5	0	95.5
Uruguay	43.3	10.2	0	53.4
Other	58.4	0	0	58.4
Total Latin America	1,321.0	66.2	94.6	2,077.9

Source: Madden *et al.* 1937, p. 77

Beginning in 1931, all Latin American borrowers except Argentina defaulted on their bond commitments, first by rescheduling the payments and later by redeeming much of the debt at a small fraction of face value. These words from an analysis during the 1930s offer a sobering reflection for the present situation:[5]

> The first defaults on foreign dollar bonds during the depression occurred in Latin America. On January 1, 1931, the Republic of Bolivia announced that . . . it was unable to pay the service on its dollar bonds. Peru . . . on March 26, 1931, . . . advised the fiscal agents for its dollar loans that it had found the 'treasury bare of funds'. . . . The Chilean government's default was announced on July 16, 1931. . . . The collapse of coffee prices, depreciation of the currency, and revolution caused the Brazilian government to announce on October 19, 1931, that it was unable to obtain the necessary foreign exchange to pay the service on its dollar bonds . . . . The Central American republics soon joined the group of defaulting countries. . . . By the end of 1933 the Argentine Government was the only [Central or] South American national government having dollar bonds outstanding which was continuing full service on all its external debts.

The difficulties of both the budgetary problem (namely raising the funds domestically) and the transfer problem caused Latin American borrowers to forego the transfer in favor of default. Ultimately, most of the bonds were redeemed at very small fractions of face value, so that the debt was erased from the books of the lenders and borrowers alike.

The OPEC crisis

By altering the rules of the game in the international oil market in 1973, the OPEC countries successfully forced a transfer of funds totalling well over $100 billion in just two years. The oil-importing countries then faced a need to transfer real resources back to OPEC countries, as the new OPEC purchasing power was utilized.

This transfer problem differed importantly from the previous ones, because in this case the main debtor country was the United States, which could pay its obligations in dollars, the universal currency for pricing oil sales. Of course, for the other oil-importing countries, the problem still required obtaining foreign currency to make payments.[6] Table 7.4 shows the size of OPEC surpluses, which represent purchasing power shifted from oil importers to the cartel's members.

The OPEC transfer problem also differs from previous ones in that several of the creditor countries do not possess the capability to absorb real resources fast enough to carry out the transfer immediately; instead, they had to accept financial claims initially for use in the future as their economies grow. While some OPEC members, such as Venezuela and Nigeria, did not face these limits on absorptive capacity, several others

The regulatory and economic environments

Table 7.4 OPEC current account surpluses, 1973–85 (US$bn)

Year	Current account
1973	4.9
1974	70.4
1975	38.4
1976	43.6
1977	27.2
1978	5.7
1979	58.5
1980	101.5
1981	40.4
1982	−11.5
1983	−19.6
1984	−4.1
1985	−10.2

Source: Estimates for 1973–6 taken from Lawrence Krause, 'For OPEC surpluses,' in Stephen Goodman (ed.) *Financing and Risk in Developing Countries*, New York: Praeger, 1978; subsequent estimates taken from Organization of Petroleum Exporting Countries, *Facts and Figures, 1985*, Vienna: OPEC, 1985.

around the Persian Gulf, such as Kuwait, Saudi Arabia, the UAE, Qatar, and Bahrain, continued to accumulate financial wealth in lieu of real purchases until the oil price decline in the 1980s. Balogh and Graham (1979) point out that this 'income effect' is a key limiting factor on the transfer which has often been ignored.

A substantial amount of the OPEC transfer problem has been eliminated by the recession of 1981–2 and the decline in oil prices during the early 1980s. At this point (1987), most of the OPEC members have achieved levels of import purchases high enough to use up the oil revenues. Many of the countries with greater absorptive capacity have even fallen into substantial trade deficits, as the value of their oil exports has fallen. In this way, the previous accumulation of dollars by OPEC members is being reduced, and the transfer is being completed now, rather than financed even longer.

The Latin American debt crisis

The crisis in the international financial system today clearly stems from the economic upheavals of the 1970s. Most importantly, the OPEC cartel's successful effort to raise oil prices caused a massive transfer of purchasing power to member countries. They, in turn, spent some of the new income on increased imports, thus partially carrying out the necessary transfer of real resources. Most of the rest of the financial wealth was invested in the international banking system, replacing deposits formerly held by industrial-country clients.

When the recession of 1974–6 left the banks with inadequate demand for their funds among traditional borrowers in the developed countries, they looked more often to non-traditional borrowers such as governments and private firms in Latin America. While these borrowers had no greater ability than before to generate income to repay international indebtedness (except for the oil exporters, Mexico, Venezuela, and Ecuador), they were offered much greater amounts of money. This fact, coupled with very low or negative real interest rates (US inflation exceeded nominal dollar interest rates for most of the 1970s), attracted the Latin American borrowers to take advantage of the new funding. Table 7.5 shows the growth of foreign borrowing by the four largest debtor countries in Latin America during the 1970s and early 1980s.

Table 7.5 Total debt of four largest Latin American countries (US$bn, year-end)

Year	Argentina	Brazil	Mexico	Venezuela
1973	6.4	13.8	8.6	4.6
1974	8.0	18.9	12.8	5.3
1975	7.9	23.3	16.9	5.7
1976	8.3	28.6	21.8	8.7
1977	9.7	35.2	27.1	12.3
1978	12.5	48.4	33.6	16.3
1979	19.0	57.4	40.8	23.7
1980	27.2	66.1	53.8	27.5
1981	35.7	75.7	67.0	29.3
8-year growth rate (annual % rate), 1973–81	24.0%	23.7%	29.3%	26.0%
1982	38.0	88.2	82.0	31.3
1983	43.6	91.2	89.3	34.0
1984	46.6	100.8	97.5	34.2
1985	50.2	102.5	98.1	35.7
1986	54.6	105.9	103.9	36.9
5-year growth rate (annual % rate), 1981–1986	8.9%	6.9%	9.2%	4.7%

Source: Adapted from William Cline 1983, Table B-1, and other material compiled by the author.

The current Latin American debt crisis is a transfer problem because the borrowers have incurred foreign-currency debt, which requires a transfer of real purchasing power to repay. It is a crisis because the borrowers do not possess adequate trade surpluses or other means of generating foreign exchange that would allow payment without severe economic dislocaton. Finally, the crisis was exacerbated by the global recession of 1981–2, which led to lower imports by Latin American exporters' main customers and more restrictions on international trade. This period also has been characterized by a return to positive, and for a while very high, real interest rates on US dollar loans — which again exacerbates the situation for the borrowers. Just as in the case of the 1930s crisis, virtually all of the Latin American countries have failed to meet their foreign currency obligations, at least temporarily, with the exceptions of Colombia, El Salvador, Guatemala, Panama, and Paraguay.[7]

Clearly, the Latin American debt crisis of the 1980s is more similar to the 1930s crisis than to the other transfer problems discussed above. For example, in the 1930s, most of the debt was initially rescheduled, as in the 1980s. Then, in the earlier period, most of the debt was eventually repaid at huge discounts on face value; which is beginning to happen today (starting with the programs endorsed by the Chilean and Mexican governments). Government policies in the 1980s have been somewhat less restrictive of trade and investment than in the 1930s, which leaves hope for increased Latin American exports, incoming foreign direct investment, and other means of effecting the transfer. Also, the US government did not force the money supply to contract during the early 1980s, in contrast to the early 1930s, so that additional negative factor did not appear. Two measures of debt-servicing capability in 1985 appear in Table 7.6

Table 7.6 Total debt-service burden for Latin American borrowers

Country	Debt-service payments/exports for 1985	Interest/ trade balance for 1985
Argentina	0.89	1.31
Bolivia	0.49	n.a.
Brazil	0.80	0.90
Chile	1.23	2.28
Colombia	0.39	8.20
Mexico	0.68	1.20
Peru	0.10*	0.37
Uruguay	0.35	2.88
Venezuela	0.49	0.59

Sources: Madden *et al.* 1937, p. 149; *International Financial Statistics*, various issues: Central Bank publications.

All of the countries listed are experiencing serious debt burdens in the mid-1980s, in which interest and principal payments consume more than one-third of export earnings in every case. Even if interest alone is measured, the burden is severe. Comparing payments to each country's trade balance in 1985 shows that most countries in the region have needed to go beyond trade surpluses to find additional foreign currency supplies that can meet interest obligations on the foreign debt.

Projections of debt-servicing problems in Latin America

One very useful way to demonstrate the immediacy of the current Latin American debt crisis is to extrapolate into the near future some key indicators of the severity of the problem. The size of total foreign indebtedness, expected interest payments, and the trade balances of the major debtor countries are measures that illuminate the situation quite clearly. While accurate estimates of these indicators are extremely difficult to make, given the fact that each one is greatly influenced by government policies as well as economic factors, at least some simple, illustrative forecasts can be made.

Using the simple autoregressive model for all variables:

$$X_t = a + bX_{t-1},$$

these indicators are estimated for 1987–91 using annual data obtained from the International Monetary Fund, the Institute for International Finance, and from central banks of the four countries. Foreign debt projections are based on the period 1982–6, when the crisis was fully realized by the lending banks. Since earlier debt growth preceded the realization of a crisis, it provides a poor basis for forecasting today. In-sample forecasts for 1985 and 1986 proved accurate to within 2 percent of the actual results that occurred. Trade balance results similarly are based only on the 1980s, since the onset of the debt crisis has put all of the countries into a continuing surplus position that appears likely to last along with the crisis. Trade balances for 1982–6 were used to generate the forecasts for 1987–91. In-sample forecasts for 1985 and 1986 proved accurate only to within 32 percent of the actual results that occurred. Finally, interest payment estimates are based on the assumption that dollar interest rates remain stable during the forecast period. Table 7.7 presents the forecasts.

The main finding from this exercise is that, under a projection of stable economic trends with no major policy or economic shifts, the crisis conditions will not improve during the next five years. Interestingly, the situation for Venezuela appears less critical than those of the other countries. With interest and fee payments averaging slightly under 10 per cent per year on the outstanding debt, Venezuela appears to have more than enough trade-generated funds to pay interest as due. Argentina appears to face the least viable situation, although Brazil with a recently

Table 7.7 Projections of Latin countries' foreign debt and exports (US$bn)

Country	1987	1988	1989	1990	1991
Argentina					
Debt	58.8	63.1	67.6	72.2	77.0
Trade balance	3.5	3.6	3.6	3.6	3.6
Interest pmts	3.5	3.8	4.0	4.3	4.6
Brazil					
Debt	107.0	107.7	108.2	108.5	108.7
Trade balance	10.1	10.5	10.7	10.8	10.9
Interest pmts	7.0	7.1	7.1	7.1	7.2
Mexico					
Debt	105.1	105.8	106.3	106.6	106.8
Trade balance	8.4	9.3	9.,5	9.5	9.5
Interest pmts	10.0	10.0	10.1	10.1	10.1
Venezuela					
Debt	37.5	38.0	38.3	38.6	38.9
Trade balance	7.6	5.9	6.5	6.3	6.4
Interest pmts	2.7	2.8	2.8	2.8	2.8

projected decline in the trade surplus for 1987 to about $5 billion, will also fall far short in foreign exchange earnings to meet debt commitments. In fact in February of 1987, Brazil's government declared a moratorium on interest payments on the foreign debt due to the trade shortfall; this problem could even be the instigating factor that leads to a major US government effort to deal conclusively with the continuing crisis that was projected above.

Considering the projections themselves, there are obviously major difficulties in placing confidence in such estimates. The implicit assumption that all relevant condition remain similar to the recent past is undoubtedly inaccurate. For one simple example, focusing on Mexico and Venezuela, any significant increase in the price of petroleum would lead to greatly increased exports — though for Brazil this situation would lead to even greater difficulties in servicing foreign debt. With varying assumptions about raw materials prices, interest rates, business cycles, and government policies in the United States and Latin America, different forecasts obviously can be generated. The simple model used above presents a clear base case for comparison, and it depicts the magnitude of the continuing crisis that exists.

Likely outcomes of the current transfer problem

How will the current transfer problem be resolved? The historical evidence shows that some form of default and ultimate write-down of loan values has been the rule rather than the exception during major crises in this century. The exception has been the OPEC-generated crisis of 1973–4, from which the industrialized oil-importing countries have emerged without defaulting. The balance-of-payments view shows that the necessary transfer can be carried out only by increasing the borrowers' trade surpluses or selling other products/services of real value (such as ownership of local companies and real estate through foreign direct investment) either now or in the future. This view also shows that the problem can be financed by new short- and long-term loans, as well as by new creation of international money (SDRs) by the IMF. In terms of real purchasing power, the burden could be diminished by an increase in US inflation which leads to negative real interest rates, thus reducing the borrowers' need to transfer real goods and services.

Partial solutions

A set of partial solutions to the crisis can be envisaged. Table 7.8 displays four categories of response that may be applied to the problem. None of the categories should be seen as the single, appropriate solution. Rather, together they form a portfolio of responses which can be used to minimize the cost of adjusting to the crisis. The first category of response characterizes the main strategy being followed at present. Creditor-country governments are not pursuing large-scale policies of new aid or lending, but instead they exhort the borrowers to try to pay and the lending banks to offer greater funding. If real dollar interests rates were to turn negative, and Latin American exports were to expand substantially, and lenders (and investors) were to gain renewed confidence in Latin America — then perhaps this do-nothing strategy could provide much of the solution. Historically, this strategy has not demonstrated much success.

Table 7.8 Responses to the Latin American debt crisis

---

1. Do nothing:
    (a)   Hope for lower or negative US real interest rates.
    (b)   Hope for renewed confidence in Latin America.
    (c)   Hope for an increase in Latin American exports.

2. Increase official lending:
    (a)   Raise the IMF's lending.
    (b)   Raise national (OECD) government lending.
    (c)   Develop new World Bank loans.

3. Form a 'debtors' cartel.'

4   Create new contractual forms:
    (a)   Allow payment in goods and services.
    (b)   Create debt–equity swaps.

A second category of response is to increase official lending from creditor-country governments and international organizations such as the IMF and the World Bank. This strategy already has been pursued to some extent, as the United States and other member governments raised their IMF quotas in 1983, thus giving the IMF more funds to lend. Additional, multilateral official financing is being proposed in the form of a possible increase in SDR allocations (i.e. increasing the international money supply). Similarly, it has been proposed that the World Bank increase its 'gearing ratio,' lending out a multiple of its paid-in capital, as do private banks. Each of these steps would raise the level of official lending to the borrowing countries — and would give the international institutions more leverage to demand monetary and fiscal reforms ('conditionality') from the borrowing governments. Another possibility, though with far less room for conditionality, is that the creditor-country governments could individually lend more funds to the debtor countries. Each of these policies would reduce the relative exposure of private-sector lenders to non-payment problems in comparison to public-sector lenders (though the majority of lending would remain in the hands of the private banks.

The third category of response is for the Latin American borrower countries to form a debtors' cartel, to force better terms from the lending institutions. This possibility has thus far been rejected by the borrowers, who do not see their interests as sufficiently similar to come to a joint policy agreement. The historical record of the 1930s shows that on that occasion, as well, no joint strategy of negotiating was followed by the borrowing countries, though virtually all of them did default on payments within a one-year period during 1931–2. Given the truly major differences in economic conditions and bargaining capabilities of these countries, the cartel concept also appears unlikely to be used in the current period.

It should be recognized that the longer the crisis persists, the more likely that 'radical' solutions such as the debtors' cartel may come into being. Notice that the cartel could pursue two substantially different strategies: it could seek less costly terms of repayment from the lending banks and official agencies, or it could press for total default (perhaps falling back on less drastic terms, once the lenders are convinced of the need to bargain). Either one of these strategies would presumably lead to a reduction in the debt-servicing burden on borrowing countries.

A final category of response is to create new contractual forms for debt repayment. These may range from the idea of accepting payment in the form of goods[8] and services instead of money, all the way to 'debt-equity swaps,' in which existing loans are exchanged for real assets in the borrowing country. The former instrument would essentially follow the lines of barter trade, placing banks in the position of intermediary *owners* of the goods being traded rather than just dealers in documents. The latter instrument enables creditor banks to sell off their Latin loans at a discount to intermediaries which, in turn, look for buyers of the discounted paper who will resell it to the issuing government for redemption (still at a discount) in local currecy.

Such debt–equity swaps are being used increasingly by creditor banks, which sell the Latin loans to other firms that seek to invest in Mexico, Brazil, and so on, and can trade the loans to issuing government for (discounted) payout in local currency. (In Mexican and Chilean external debt, the swap market has developed substantially during 1986-7, to over US$1 billion.) This type of instrument offers an important vehicle for alleviating the severity of the debt crisis. A limiting factor to be kept in mind is that the amount of money loaned by US money center banks is so large in relation to their capital bases, that these major lenders cannot sell more than a small portion of such debt very quickly. Consequently, the Latin American countries cannot eliminate more than a fraction of their foreign liabilities through this vehicle at any one time.

These swap agreements possess the attribute that they can carry out the process of reducing the debt burden without requiring new governmental agencies or accords. They function in the private sector via negotiations with the debtor governments that lead to a discounted, local-currency loan payback which is below a loan's original face value. These swaps involve original lending banks with market-making intermediaries that themselves may be banks or investment firms, which in turn sell discounted loans to other banks or firms that want local currency. Discounts in this swap market in recent years were as shown in Table 7.9.[9] Notice that trading in loans owed by governments in the four countries studied here ranges from a low of 56 per cent of face value for Mexican debt up to 83 percent of face value for Colombian debt in 1986, with values dropping continuously since then.

Table 7.9 Discount prices for Latin American debt (%)

Country	*July 1986*	Discount relative to face value of loan *April 1987*	*February 1988*
Argentina	66	58–60	30
Brazil	76	63–6	45
Chile	67	68–90	60
Colombia	83	84–6	63
Ecuador	65	54–8	32
Mexico	56	57–60	48
Peru	n.a.	14–18	7
Venezuela	75	73–5	55

Source: Shearson Lehman Brothers International, Inc.

(Along a similar line, the World Bank could function as an intermediary between countries and foreign commercial banks, buying loans at a discount from the private banks and issuing replacement loans at better terms to the borrowing countries. This scheme faces the same limitations as the previous one, but it places the World Bank in the position of market-maker rather than leaving that function to the private sector.[10])

None of these four responses to the Latin American debt crisis is likely to be a sufficient solution by itself. Each of them has contributed to avoiding a financial panic during the five years that the problem has been recognized.

The same problem from a new perspective

A really final solution to the debt crisis requires the re-establishment of confidence in Latin American economies sufficient to attract domestic and foreign investors to resume investing their funds locally. To some degree this is already happening, as almost all of the countries in the region except Ecuador, Mexico, and Venezuela (the main oil exporters) have had positive economic growth in 1985 and 1986. In fact, the crisis today is not that the debt is threatening overall economic development in Latin America, but that it threatens to cloud the view of foreign suppliers of funds who could increase the rate of investment there.

The debt should be viewed in three categories: existing loans to governments and private borrowers who have been disallowed normal access to dollars to pay foreign obligations; new loans to the private sector for exports, manufacturing, and so on; and new loans to governments. The existing foreign debt that has been restructured by each Latin American country during 1982–7 must be settled using the methods discussed in the previous section. There is no reason to expect full elimination of these obligations any time in the next few years — though with more renegotiations, debt–equity swaps, inflation, increased trade, and good government management of economic policies, the existing debt can be handled without major economic or financial crises in Latin America or in the rest of the world.

The second part of foreign debt in Latin America is new financing of private–sector activities., Export and import finance have increased significantly in 1986 and 1987. US and other foreign commercial banks are demonstrating a willingness to lend to creditworthy clients in the region, though with more collateral and more use of EXIMBANK-type guarantees than before. Lending to Latin American firms for local production and sales is lagging the international sector's growth, understandably since the access to dollars still remains in question in several countries. It is asserted that this type of financing, too, will resume an upward trend as the region continues to demonstrate positive economic growth. Basically, new private-sector lending is escaping from the 'debt overhang'.

What remains in serious question is the access of Latin American governments to foreign commercial bank credit. Having lost or still expecting to lose substantial sums to these borrowers, how can the banks be expected to resume lending to the governments at this point? From one perspective, the commercial banks should be focusing their loans on private business anyway, so a decline in government lending should be welcomed. From another perspective, as long as a government can demonstrate its creditworthiness through proposals for projects that will generate dollar cash flows or even the use of collateral such as government dollar deposits in the foreign banks or other credible instruments, then governments can regain some access to foreign commercial bank financing. The use of collateral in government loans would be a new twist in cross-border lending, but it is far from a 'radical' solution.

By considering the debt in these three categories, it can be seen that even with the very large overhang of existing foreign debt, the prospects for Latin American economic growth and even increased (voluntary) foreign lending are more positive than the initial analysis implied. While government borrowing may very well return to a much greater dependence on foreign official financing (and a lower rate of growth), private-sector borrowing already has recovered substantially from the early 1980s.

Conclusions

The transfer problem of Latin American debt in the last few years has many parallels with those countries' problems during the Depression of the 1930s. That earlier debt crisis ended with US lenders essentially accepting almost total default on the loans (bonds) and a virtual end to new loans to that region until the Second World War. In the late 1980s, Latin American borrowers face even greater debt-servicing burdens relative to their current export flows; but the industrialized countries are now out of recession, so that demand for Latin American exports and willingness to invest in the region have not fallen nearly as far as in the earlier period.

The need to transfer real purchasing power still remains in the present crisis. History offers a bleak view of the expected outcome, though perhaps it shows the policies that will *not* work. Restriction of imports and reduction of money supply growth did not work well in the 1930s. The political willingness of the borrowing governments to keep trying to pay can be encouraged today by industrial-country policies which stimulate Latin American exports, foreign investment in the region, and agreements by lenders to reduce their charges for money on loan. In addition, the relatively orderly process that has arisen for selling individual loans at a discount through intermediaries that are willing to accept local currency instead of dollars provides another safety valve for alleviating the crisis. And finally, if government borrowers can agree to some form of collateralized borrowing as is occurring in private-sector loans, then renewed lending for government projects also may take place. The net result of these conclusions is that the transfer probably will be partially carried out through trade surpluses and foreign direct investment, partially financed for an indefinite period, partially inflated away, and partially renounced by the borrowers (especially through *debt–equity* swaps) — very similar to the experience with German *reparations and Latin American* debt in the 1930s.

Notes

*This article is reprinted with permission from *The World Economy* (Fall 1988).
1. Robert Grosse, 'The transfer problems and Latin American debt', University of Miami Graduate School of International Studies *Occasional Paper 1985-3*, 1985.
2. The Marshall Plan (1948–52) unilaterally transferred approximately $13.15 billion dollars from the United States to Western European countries. About two-thirds of these funds were used to buy goods and services from US suppliers, and the rest for similar purchases from non-US suppliers. The Marshall Plan does *not* constitute a transfer problem, because the payments were made in *domestic* currency rather than foreign exchange. It may be considered a *budgetary* problem.
3. J.M. Keynes, 'The German transfer problem', *The Economic Journal*, March 1929, p. 1.
4. John Madden, Marcus Nadler, and Harry Sauvain, *America's Experience as a Creditor Nation*, New York: Prentice-Hall, 1937, p. 108.
5. Ibid., pp. 11–14.
6. OPEC members did discuss the possibility of pricing oil in yen or marks, as the dollar's value fell during 1974–6, but this strategy was not followed.
7. Robert Grosse, 'The transfer problem and Latin American debt', University of Miami Graduate School of International Studies *Occasional Paper 1985-3*, 1985. During 1987–8, the other countries also have renegotiated their foreign official debts.
8. Allan Meltzer has suggested that Latin American countries sell partial ownership of government-owned companies to pay part of the debt: Allan Meltzer, 'A way to defuse the world debt bomb', *Fortune*, 28 November 1983.
9. These loan discounts primarily reflect swaps of loans due from different Latin American borrowers between US commercial banks, rather than debt-equity swaps. In the past year, the debt–equity swaps have become a much more important part of this swap market.
10. See, for example, Richard Dale and Richard Mattione, *Managing Global Debt*, Washington, DC: Brookings Institution, 1983, pp. 42–8.

Bibliography

Balogh, Thomas, and Andrew Graham (1979) 'The transfer problem revisited: analogies between the reparations payments of the 1920s and the problems of the OPEC surplus,' *Oxford Bulletin of Economics and Statistics*, August.

Cline, William (1983) *International Debt and the Stability of the World Economy*, Washington, DC: Institute for International Economics.

Dale, Richard and Richard Mattione (1983) *Managing Global Debt*, Washington, DC: Brookings Institution

Grosse, Robert (1984) 'Solutions to the Latin American debt crisis,' in John Disbury (ed.) *Strategies for Dealing with the Future*, Bethesda, Maryland: World Future Society.

Grosse, Robert (1985) 'The transfer problem and Latin American debt', *Occasional Paper 85-3*, University of Miami Graduate School of International Studies.

Johnson, Harry G. (1976) 'Notes on the classical transfer problem', *Manchester School of Economics and Social Studies*, September.

Keynes, J.M. (1929) 'The German transfer problem', *Economic Journal*, March.

Kindleberger, Charles (1973) *The World in Depression*, Berkeley: University of California Press.

Madden, John, Marcus Nadler and Harry Sauvain (1937) *America's Experience as a Creditor Nation*, New York: Prentice-Hall.

Meltzer, Allan, 'A way to defuse the world debt bomb,' *Fortune* 28 November.

Metzler, Lloyd (1942) 'The transfer problem reconsidered,' *Journal of Political Economy* June.

Ohlin, Bertil (1929) 'The reparation problem: a discussion 1', *The Economic Journal*, June; and 'Mr Keynes' views on the transfer problem II,' *The Economic Journal*, September.

Samuelson, Paul (1968) 'The transfer problem and transport costs,' in Richard Caves and Harry Johnson (eds) *Readings in International Economics*, Homewood; Ill: Richard D. Irwin, pp. 115–47.

Part three
Corporate Strategies

Chapter eight

Competitive advantages and MNEs:
A Latin American application

Strategies for multinational enterprises (MNEs) have been a favorite topic for management analysts during much of the period since the Second World War. One early view saw MNEs as exporters forced by tariff barriers and other host-country restrictions to act defensively, using foreign investment to protect their foreign markets (Barlow and Wender 1955). Another saw MNEs as possessing some strengths relative to local firms (such as economies of scale in production and advanced technology) which enabled them to succeed in foreign markets that they knew relatively little about (Hymer, 1960). Yet another view developed the idea of an international product cycle based on companies' development of new technologies and marketing know-how, that leads from introduction of a new product in one country, to its export and local production over time in other countries (Vernon 1966). Recently, many analysts have been examining the issue of competitive strategy (Porter 1980, 1985), which posits key strategic choices for companies — such as the choice between following a strategy of product differentiation versus low-cost, low-price competition, versus focus on a specific market niche — based on the company's competitive strengths, such as proprietary technology or a well-known brand name. The present chapter offers a detailed framework for systematically incorporating much of this thinking in multinational firms.

Three central questions are addressed:

1. What characteristics give MNEs the ability to compete in the various national markets in which they do business?
2. Which competitive advantages are associated with which industries or groups of firms?
3. How does company performance relate to these advantages?

Each of these questions is considered in the context of Latin America in the mid-1980s.

The first section of the chapter presents a framework for examining competitive advantages that a multinational firm may obtain to enable it to compete with other MNEs and with local firms in each national market it enters. The second section describes the sample of firms that was used to explore the importance of competitive advantages and the methodology of analysis. The third section empirically shows which competitive advantages are associated with different kinds of firms. The fourth section looks at the relationship between company characteristics and performance; and the final section gives some conclusions.

The concept of competitive advantage

Competitive strategies have changed dramatically since the advent of large numbers of multinational manufacturing firms in the 1940s and 1950s. Today, to be competitive in almost any substantial market that is not government-controlled, a firm must develop strengths that enable it to survive against foreign as well as domestic rivals. By looking carefully at those specific elements of competitiveness that relate to *multinational* firms, the present analysis establishes some bases for competitive strategy in these firms.

Competitive advantages are the skills or abilities that a company possesses that allow it to earn above-average profits in competition with other firms. Such abilities arise from the company's functioning in various markets — for example buying inputs for production and selling outputs to customers — and from the company's internal operation, for example managing blue- and white-collar workers and manufacturing or providing a service. In each instance, the company that achieves an advantage has lowered its costs or increased its revenues (or reduced its risk) relative to competitors, for example through economies of scale, more effective promotion or better-diversified activities.

Figure 8.1 presents a view of the relationships between major categories of a company's internal and external concerns. Using this scheme, competitive advantages can be categorized as those deriving from company strategy, those deriving from intra-company structure, and those deriving from environmental factors such as government policies, inter-firm rivalry, and geographic and cultural characteristics. In other words, the competitive advantages arise from the functioning of the firm's own 'internal market' (i.e. internal purchases and sales, labor contracts, technology transfer from one affiliate to another, etc.) and the external markets in which the firm participates.

For example, Beatrice Companies has discovered that it can successfully transfer product differentiation strategies developed in one country (e.g. innovative cookies in Brazil) to other countries (e.g. Venezuela) in which it sells similar foods; the company has managed its internal resources to obtain an economy of scale in advertising and physical distribution for its differentiated products. Westinghouse Corporation

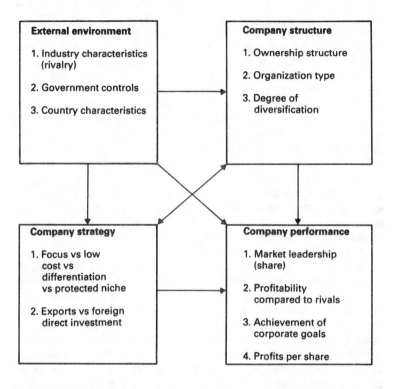

Figure 8.1 Relations between environment, strategy, structure, and performance (each factor suggests a source of competitive advantage)

Note: This figure illustrates several important relationships between characteristics of a company and its environment that can be used to create superior performance. For example, a focus strategy can be used to target some market that involves only the government as a buyer, and the MNE as seller. Success in following this strategy will give the company a protected market niche, which should lead in turn to above-average profits.

has demonstrated an ability to negotiate contracts to provide major power-generating facilities with governments in less-developed countries and in Eastern Europe; it now possesses a competitive advantage in government contracting and a protected business position in several national markets. Hoover, Caterpillar Tractor, and many other firms have developed long track records of delivering high-quality products and superior after-sale service; this is a kind of competitive advantage in goodwill, since it results in greater demand for their products than those of rival firms. IBM and many large Japanese companies have commitments to permanent employment that are passed on to employees and substantially followed by the companies' executives — leading to better internal labor and personnel relations than in many other firms. L.M. Ericsson, the Swedish telecommunications manufacturer, has developed small electronic switching systems that suit many customers who need less-extensive, and less-costly telephone systems than those produced and sold by its major rivals in international business: Ericsson has developed a market niche advantage. Each of these examples of competitive advantages gives the company the ability to out-compete its rivals in a particular business, and most of them have been discussed separately elsewhere in the literature.[1]

The managerial question that remains after all of these examples is: how can potential advantages be determined so that the key ones can be selected and utilized by a firm? The purpose of this section is to create a framework for managers to think about such advantages and to understand how they will aid a company.

The basic framework, which is used fairly widely by business analysts, views companies as organizations that carry out a set of economic activities. Let us define the economic activity of the firm as essentially production, subdivided into obtaining of necessary inputs, actual producing of the goods/services, their distribution to points of sale, and marketing to the customers.[2] Figure 8.2 presents a sketch of the basic aspects of production in which competitive advantages may arise.

Purchasing is the first stage, in which inputs are assembled for production of the product or service. Then, either in that location or after transferring the inputs to another location(s), basic production is carried out. If additional assembly is needed, that may be done in the same location or distributed to one or more assembly or finishing facilities. Finally, the product or service is marketed to customers in the same location and/or distributed to others.

Notice that at each step of the production process the firm must choose combinations of inputs — manpower, physical capital, financial capital, and information — that will generate the output; it must decide how to organize them; and it must decide how to deal with the risks involved, for example through simply bearing the risks or seeking to transfer them to other firms. Managers should make these decisions to minimize the cost of completing the economic activity. So, for example, if it is less expensive to buy components than to produce them internally, the firm should contract out for the components. Similarly, if the firm chooses to bear the risk of losses, rather than contract with an outside insurance provider to cover that possibility, it is providing 'self-insurance' (which should have a lower cost than the outside insurance). In all, the firm can be viewed as a value-maximizing organization which decides on optimal combinations of inputs at each stage of the production process to generate its output and thus profit.

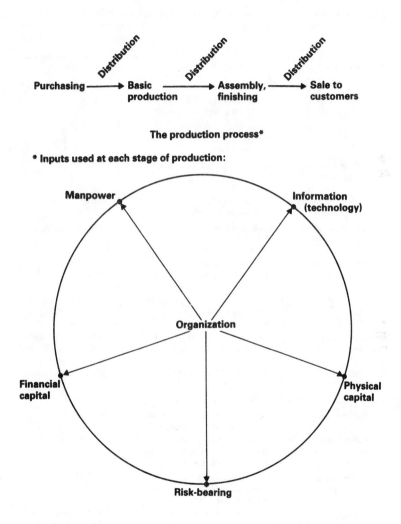

Figure 8.2 Sources of competitive advantages in the production process

Two key points here are that (1) the firm must attempt to evaluate the impact on profits of not only its own activities, but also those of its rivals in the market and those of other outside actors (especially governments), and (2) the firm must consider the functioning of its own internal

Table 8.1 Selected competitive advantages of multinational enterprises

Advantage	Description	Example of product or company
1. Proprietary technology	Product or process technology held by a firm that others can obtain only thru R&D or contracting with the possessor	IBM; DuPont; Johnson & Johnson; AT & T
2. Scale economies in production	Large-scale production facilities that lower unit costs of production	US Steel; Ford; Boeing
3. Scale economies in purchasing	Lower unit costs of inputs through purchasing large quantities	Any very large firm
4. Scale economies in financing	Access to funds at a lower cost for larger firms	AT&T; IBM; EXXON
5. Scale economies in distribution	Sales operations in several countries, allowing a firm to serve a 'portfolio' of markets	Any very large firm
6. Scale economies in advertising	Sales in several countries allowing somewhat standardized advertising	Coca-Cola; IBM; Nestlé
7. Government protection	Free or preferential access to a market limited by government fiat	General Dynamics; Renault; British Petroleum
8. Goodwill based brand or trade name	Reputation for quality, service, etc., developed through experience	RCA; Levis; Sears; Hoover
9. Superior location	Better access to the market	Restaurants; hotels
INTERNATIONAL COMPETITIVE ADVANTAGES		
10. Multinational market access	Knowledge of and access to markets in several countries	Any MNE
11. Multinational sourcing capability	Reliable access to raw materials, intermediate goods, etc., that reduce costs	Any MNE
12. Multinational diversification	Operations in several countries so country risk and business risk are reduced	Any MNE
13. Managerial experience in several countries	Skill for managing multi-country operations gained thru experience in different countries	MNEs with managers experienced in international business

market, including the 'people problems' of labor/personnel relations that must be solved to achieve optimal outputs. From this scheme, we can derive the main competitive advantages that have been noted elsewhere (e.g. in Porter 1980, 1985; Grosse 1985), and we can see many additional areas in which advantages may arise. Competitive advantages at each stage of the production process are considered separately below. Table 8.1 depicts a selected number of them, divided into those accessible to firms in general and those available only to multinational enterprises.

Note that the most commonly discussed advantages, proprietary technology and production economies of scale, appear on the list, but so also do many other kinds of scale economy that reduce costs, as well as 'goodwill' (often recognized through a brand or trade name) and government protection that can enable the firm to escape competition because of preferential government policy. Among the specifically international advantages are the revenue-increasing capacity of a firm with marketing operations in several countries, the cost-decreasing capacity of a firm with sourcing operations in several countries, and the risk-reducing capacity of a firm with business in several nations (and thus subject to different laws, social, political and economic conditions.) A final international competitive advantage that often exists for firms whose managers have relatively great experience in international business is a managerial advantage due to these managers' use of this information in their decision-making.

Methodology and sample

The rest of this chapter is concerned with exploring the view of competitive strategy presented above in the context of Latin American countries during the mid-1980s. In this section, the methodology of analysis is discussed, then the sample and sampling procedure are characterized.

Methodology

In order to investigate the importance of competitive advantages to MNEs, and to answer the three research questions, a sample of multinational firms was needed. This sample was drawn entirely from large, US-based firms — namely members of the most recent Fortune 500 list. Thus conclusions drawn from the analysis can be applied to this particular class of firms and only with caution to others.

A questionnaire was developed to request information from company executives with responsibility for Latin American operations in each firm. This questionnaire was pre-tested in interviews with MNE managers

161

working in South Florida at the Latin American headquarters or marketing offices of half a dozen US firms. The finalized questionnaire was composed of thirteen closed-ended questions covering company characteristics such as size, structure, age, and performance measures, as well as one question covering a list of sixteen competitive advantages. The list of competitive advantages was developed from the business strategy literature and from discussions with company managers. The question about competitive advantages asked only for managers' opinions, rather than factual, recorded data. Additional information about some advantages was available from other, published sources.

Responses to the survey were tabulated to obtain frequency distributions on each question and for correlations among firm type, competitive advantages, and performance. Multiple regression analysis was used to explain performance results based on company characteristics and advantages. In addition, factor analysis was used to try to group some of the competitive advantages into meaningful clusters.

The sample

The sample was drawn from the 1986 listing of the Fortune 500 US-based industrial corporations. First, an effort was made to include firms that truly have a Latin American strategy. Only those firms with affiliates in at least two of the largest four Latin American countries (namely Argentina, Brazil, Mexico, and Venezuela) were considered. This resulted in an initial universe of 165 firms.

Second, an initial mailing of the questionnaire was made in July of 1986. The survey instrument was sent to the Latin American division manager at each firm, or to the head of the international division if no Latin American manager could be identified. About twenty of these executives were located at offices in South Florida, and the rest at the home office or in Latin America. About forty responses were received, including several that simply replied that the firm would not participate in the study. A second mailing to non-respondents was carried out in August 1986. In all, over seventy responses were received, yielding fifty-six usable questionnaires.

Empirical evidence: competitive advantages in US MNEs operating in Latin America

Managers' views of competitive advantages

The survey demonstrated a broad agreement by MNE managers across industries and firm types concerning their competitive advantages. They uniformly pointed to competitive strengths based on proprietary technology and on company/product reputation (based on brand names or company name). Table 8.2 presents the mean scores for each of the sixteen competitive advantages and their standard deviations. Note that clearly the most important strengths as viewed by these managers were proprietary technology and goodwill based on brand name or company name. Also rated highly were the possession of an international marketing network and managerial experience in Latin America. The mean scores of the top two competitive advantages (goodwill and proprietary technology) were significantly higher than any of the others, according to simple t-tests on their mean values.

A second broad measure of the relative importance of these competitive advantages was carried out through factor analysis. The analysis

Table 8.2 Competitive advantages of US MNEs in Latin America (based on survey responses of 56 firms)

Advantage	Mean score*	Standard deviation
1. Goodwill based on brand name or firm name	1.24	0.69
2. Proprietary technology	1.13	0.65
3. Managerial experience in Latin America	0.71	0.87
4. International marketing network	0.69	0.74
5. Financial capability	0.63	0.85
6. Knowledge of local conditions	0.55	0.67
7. Worldwide sourcing capability	0.53	0.68
8. Production economies of scale	0.52	0.81
9. Human resource management	0.51	0.67
10. Size of firm relative to rivals	0.50	0.90
11. Distribution economies of scale	0.46	0.94
12. Low costs — low wages	0.22	0.69
13. Low costs — low salaries	0.19	0.67
14. Better labor productivity	0.10	0.71
15. Government protection	0.06	1.00
16. Access to scarce raw material	0.04	0.39

* Scores range from 2 = major competitive advantage to 0 = neutral to −2 = major competitive disadvantage on a five point scale.

yielded six factors with eigenvalues greater than 1. The Varimax rotated factor loadings for the sixteen variables are presented in Figure 8.3a. These six factors accounted for 70 percent of the retained variation in the survey responses concerning competitive advantages.

Figure 8.3a Clusters generated by factor analysis of sixteen competitive advantages

Factor 1: SCALE ECONOMIES
Economies of scale in production and distribution;
relative firm size; access to financial resources

Factor 2: LOW-COST LABOR
Wages and salaries

Factor 3: 'PEOPLE SKILLS'
Proprietary technology; managerial experience;
knowledge of local business conditions;
human resource management

Factor 4: TRADE RELATED VARIABLES
Government protection; international marketing network

Factor 5: MARKET or SOURCING POWER
Goodwill based on brand or company name;
access to scarce raw materials

Factor 6: PRODUCTION EFFICIENCY
World-wide sourcing of production;
labor productivity

Interpretations of the six factors are presented in Figure 8.3b. Notice that Factors 1, 2, and 6 group three kinds of cost-reducing variables — namely economies of scale, low compensation levels, and production efficiencies. Factor 2 groups human resource skills that are embodied in the work force such as technological knowledge, managerial experience, and human resource management. The other two factors are more difficult to interpret, though their loadings are equally strong.

Factor 4 groups government protection and possession of an international marketing network. These two factors may be linked for firms that depend importantly on transhipping products within their international marketing networks, and which also view government protectionism as an important constraint on these activities. Presumably, the possession of an international marketing network would be a competitive advantage, while these (foreign) firms viewed government protection as an advantage of their local rivals and a disadvantage to

Figure 8.3b Varimax Rotated Factor Loadings

Competitive advantage	Factor 1	Factor 2	Factor 3	Factor 4	Factor 5	Factor 6	Communality
Proprietary technology	−0.18	0.07	0.58	−0.05	−0.41	−0.25	0.6104
Goodwill	0.07	−0.13	0.49	0.28	0.58	−0.10	0.6848
Production scale economies	0.63	0.08	−0.15	−0.47	−0.05	−0.30	0.7429
Distribution scale economies	0.73	−0.06	−0.10	0.01	0.08	−0.48	0.7798
Access to raw materials	0.05	0.22	−0.13	−0.10	0.82	−0.03	0.7522
Government protection	−0.12	0.04	−0.06	−0.84	−0.10	−0.12	0.7491
Managerial experience	0.16	0.15	0.80	−0.08	−0.11	0.07	0.7056
Relative firm size	0.81	0.19	0.06	0.20	0.14	0.06	0.7533
Financial capability	0.71	−0.05	0.31	0.09	−0.02	0.01	0.6158
World-wide sourcing	0.03	−0.20	0.08	0.05	0.17	−0.75	0.6445
Labor productivity	0.20	−0.16	0.25	−0.06	−0.14	−0.66	0.5864
Low wages	−0.09	−0.86	−0.12	−0.01	−0.05	−0.33	0.8794
Low salaries	−0.01	−0.90	−0.01	−0.06	−0.10	−0.04	0.8254
Int'l dist. network	0.09	0.23	0.00	0.66	−0.18	−0.40	0.6989
Local knowledge	0.39	−0.30	0.60	0.12	0.00	−0.04	0.6154
Human resource management	−0.05	0.08	0.63	0.18	0.21	−0.26	0.5436
EIGENVALUES	3.20	2.26	1.98	1.48	1.15	1.12	
% VARIATION EXPLAINED	20.01	14.13	12.36	9.22	7.21	7.00	
CUMULATIVE % EXPLAINED	20.01	34.14	46.50	55.72	62.92	69.92	

themselves. This interpretation is intuitively sensible, though the reasons that the two variables loaded together are not obvious.

Factor 5 groups goodwill based on brand name or trademark and access to scarce raw materials. Both of these variables provide the possessor firm with market power relative to rival firms, but the two types of power are quite different (One on the demand side, and one on the supply side.).

Each of these factors was subsequently used in the attempt to explain company performance, as discussed below (pp. 170ff).

Company and industry differences in competitive advantages

It was expected that different kinds of firms (e.g. high-tech vs. low-tech) and different industries would demonstrate different competitive advantages. Thus the next step was to compare advantages among firm types, to distinguish company and industry characteristics. Five characteristics were used to compare firms:

1. High-tech versus low-tech.
2. Consumer versus industrial products.
3. High-advertising versus low-advertising firms.
4. New versus experienced firms in the region.
5. Joint-ventures versus wholly-owned subsidiaries.

A priori reasoning was that high-tech firms would view technology as a key advantage; marketing-intensive firms would view goodwill as a key advantage; and industrial-products firms would view production economies of scale as very important. No other relationships were hypothesized, ex ante. Cross-tabulations on the range of five dichotomies of firm types compared with competitive advantages were used to distinguish the differences. Table 8.3 shows the significant ($\alpha < 0.10$) correlations for each firm type. Notice that very few of the competitive advantages correlated highly with differences in firm type; that is, the advantages were viewed quite similarly across company categories. The few differences include the following:

1. *High-tech firms* view world-wide sourcing as a more important advantage.

   High-tech firms view production scale economies as a less important advantage.

2. *Consumer products firms* view goodwill as a more important advantage.

   Consumer products firms view larger firm size as a more important advantage.

3. *Newer firms* gain greater advantage from an international marketing network.

   Newer firms view low wages as a more important advantage.

   New firms view labor productivity as a more important advantage.

4. *High-advertising firms* view goodwill as a more important advantage.

   High-advertising firms view government protection as a less important advantage.

Several of these results need no explanation, though a number of them are not intuitively clear. For example, high-tech firms appear to see world-wide sourcing as a very important competitive advantage, and none of the other advantages as similarly important (not even technology itself, which is also viewed as important by the relatively low-tech firms). Both classes of firms that are expected to be 'marketing-intensive' showed the expected dependence on goodwill as a key competitive advantage.

Interestingly, both low-tech firms and firms recently established in the region (since 1970) viewed low wage costs as an important advantage relative to their competitors (some of whom may be importers of products from higher-wage countries).

Table 8.3 Significance of competitive advantages to different firm groups

Company types: competitive advantages	Measure	
High-tech vs low-tech:	(Above or below US industry average)	
World-wide sourcing	Greater technical intensity — greater sourcing	p=0.058
Production scale economies	Greater technical intensity — lower scale economies	p=0.065
Consumer vs industrial products:	(SIC codes for two key products)	
Goodwill	Consumer products — greater goodwill	p=0.011
Relative firm size	Consumer products — greater firm size	p=0.046
New vs experienced firm:	(Year of entry of first Latin American affiliate)	
Int'l. mktg network:	Newer firm — more use of int'l. mktg network	p=0.007
Low wages	Newer firm — greater advantage of low wages	p=0.033
Labor productivity	Newer firm — greater labor productivity	p=0.064
High-advertising vs low advertising:	(Advertising expense ratio to sales)	
Goodwill	Higher advertising — greater goodwill advantage	p=0.006
Government protection	Less advertising — more govt protection	p=0.010
Joint venture vs wholly-owned subsidiary	Nothing significantly different	

These results are instructive in showing what company managers feel to be their important competitive strengths in competition in Latin America. We are interested also in establishing links between the company characteristics and actual performance. The next two sections explore this issue.

Taking advantage of competitive advantages

The conceptual base

The final step for the manager in creating a competitive strategy is to determine how to take advantage of the firm's accumulated competitive advantages and minimize the impact of its competitive disadvantages. The solution depends, of course, on the specific set of advantages obtained by the firm and the competitive conditions in the industry (or industries) and countries in which it competes. A natural resource company generally will be very limited in its ability to own mines, wells, land, or other sources of raw materials in most countries of the world today. Instead, the resources often must be purchased through long-term contracts or in spot markets — and a key advantage may accrue to firms that have multiple suppliers. Manufacturers of mature industrial products may need to reduce costs by seeking low-cost assembly locations, perhaps in newly-industrializing countries such as Mexico, Taiwan, or Malaysia. Producers of military goods may need to produce locally some of their components and/or products, to comply with government restrictions in the purchasing countries, regardless of cost conditions that favor other sites.

An array of vehicles for doing business in various national markets is sketched in Figure 8.4, which also suggests several criteria for choosing among the alternative methods of exploiting competitive advantages.

Figure 8.4 Methods of exploiting competitive advantages through foreign involvement

FORM OF FOREIGN INVOLVEMENT	Added income	Capital commitment	Management commitment	Technology commitment	Political risk	Flexibility	Impact on rivals
			DECISION CRITERIA				
EXPORTS	?	Low	Low	Low	Low	High	?
examples	(a) Direct						
	(b) Through a distributor						
	(c) Through a trading company						
CONTRACTING	?	Low	Possibly high	Possibly high	Low	?	?
examples	(a) Licensing technology						
	(b) Franchising						
	(c) Management contracting						
PARTIALLY-OWNED DIRECT INVESTMENT	?	?	Possibly high	High	Medium	Low	?
examples	(a) Joint venture with local company						
	(b) Joint venture with foreign company						
	(c) Joint venture with government						
WHOLLY-OWNED DIRECT INVESTMENT	?	High	High	Low	High	?	?
examples	(a) Assembly plant for local sales						
	(b) Basic manufacturing						
	(c) Raw materials' extraction						
	(d) Offshore assembly plant						

Note: Each column represents a dimension for decision-making that should be considered when choosing a method to use in exploiting a competitive advantage. The rankings will differ from company to company and also across countries; the entries in the table are for illustrative purposes only.

Notice that the 'bottom line' is not a simple cash-flow analysis of market-serving alternatives, but rather it is a strategic evaluation of how the method fits into the firm's full global organization and strategy. For example, a project in Japan whose net present value is lower than an alternative project in France may be accepted, if the impact of the project in Japan is to reduce the profitability of a competitor in Japan and thus limit the competitor's ability to compete in other markets as well.[3] Similarly, a project with a lower net present value than some other alternative should be chosen if the more profitable project demands too much of the firm's scarce management time or if it presents too great an exposure to economic, political, or foreign exchange risk. In all, market-serving alternatives should be chosen on the basis of their total impact on the global corporation, and not just on their expected returns as independent activities.

The multinational firm may take advantage of its competitive strengths through multiple vehicles in various markets; ownership of a factory or office in each country is unnecessary. Hilton Hotels earns a substantial return on its business activities in different countries, even though the majority of hotels in the chain world-wide are franchised out to independent hotel owners. Similarly, pharmaceuticals manufacturers routinely contract with other firms in the industry to produce their drugs under 'tolling' agreements, which are essentially licenses to manufacture ethical drugs using the proprietor firm's name. The Dow Chemical pharmaceuticals plant in Peru formulates drugs for Merck, Eli Lilly, and other multinational drug companies; in other countries, Dow contracts out to other firms to produce its products in the same manner. The correct strategy depends on the entire competitive position of the company in all of its markets.

Overall, the firm needs to take stock of its own competitive advantages, plus those of its competitors, and select a strategy that emphasizes its strengths and minimizes the impact of its weaknesses. The multinational firm must consider the strategies of its competitors not only in its home market but also in every relevant foreign market. Finally, managers must realize that no matter how strong the competitive advantage, other firms are likely to chip away at it, and ultimately the advantage will be gone;[4] to be competitive in the long run will require regular review of the competitive position of the firm in its various markets and products, and then adjustment of the strategy to take advantage of environmental and internal changes.

Examples of MNEs' competitive strategies in Latin America

The firms in the sample were not questioned about their broad competitive strategies in Latin America. With our focus on competitive advantages, industry types, and performance, the issue of integrating Latin American activities into the MNE's total business was not considered directly. Some specific examples of competitive strategies of firms in the sample can, however, be sketched.

For example, it is quite clear that the chemicals firms (see also Ch. 11 below.) centralize their R & D in industrial countries and their production of basic chemicals mostly in industrial countries (with several exceptions in Brazil and Mexico). On the other hand, they have widespread sales offices in Latin America and also fairly numerous final processing facilities. The firms are usually considered technology-intensive, though their major products in Latin America sometimes are commodity chemicals. Very few of the firms export significant amounts of their products within Latin America, while most of them do import into the region from their North American and European plants. In this high-tech industry, production is generally centralized outside of the region, with local processing plants, sales offices, and distributors as the most common forms of market servicing.

Even less local production takes place in high-tech industries such as computers,[5] aircraft, electronic instruments, and telecommunications equipment. These industries are characterized by even more centralization of production in the industrial countries and only limited local assembly in Latin American markets.

Industries that are not dominated so importantly by advanced technology show much greater amounts of direct investment in local production within Latin America. Food-processing firms often have many manufacturing subsidiaries spread throughout the region, as do producers of electrical appliances, tires, and other relatively low-tech products. The firms in the sample tend to follow these production strategies quite regularly.

The final question to be answered with the information gathered in the sampling of US MNEs relates to their performance in Latin America. Using the competitive advantages discussed previously and following production strategies as noted above, how have the firms performed in their Latin American business?

Empirical performance

Performance results for the entire sample

The basic measure of performance used in this analysis was market share of the firm's main product. While far from a perfect measure of earnings performance, this indicator has the desirable characteristic that it is not subject to bias due to the firm's internal pricing strategy (that may bias profits toward or away from the Latin American affiliates) or due to other factors such as shifting exchange rates or remittance policies. Market share was averaged for all of the countries in which the firm operates in Latin America, yielding one aggregate measure.

A second measure of performance, the company manager's opinion on how his firm was performing relative to its main rivals in the two main products, was collected as well. The manager's opinion on this performance in each of the four countries was requested, then the responses for that company were pooled to create a continuous distribution. Regressions on this dependent variable also are presented below.

The market share measures for the entire sample were regressed against the various competitive advantages using hypothesized relationships as well as using the advantages scored as most significant by the company managers. Table 8.4 shows results of these regressions. As may be expected, those two competitive advantages that were most often cited also were most significant as factors explaining performance. Proprietary product or process technology and goodwill based on a brand name or the firm's name explained about 3/4 of the variance in performance (i.e., market share) in a variety of specifications of the relationship.

Only one other variable, the length of the firm's experience at operating in Latin America, added significantly to the explanatory power of the regression. Apparently, longer experience in Latin America does translate into superior performance for these firms.

Table 8.4 Performance relationships with competitive advantages

Performance measure	Competitive advantage 1	Competitive advantage 2	Competitive advantage 3	R^2
Average market share	Technology (+4.05)	Goodwill (3.21)	n.r.	0.71
Average market share	Technology (+3.92)	Advertising (+4.41)	n.r.	0.79
Average market share	Technology (+2.37)	Advertising (+2.24)	Age of subs. (+2.35)	0.80
Average market share II	Technology (+3.24)	Goodwill (+3.55)	n.r.	0.72

The analysis of managers' opinions of their firms' relative performance compared to rivals resulted in similar correlations. Only proprietary technology remained significant as an explanatory variable in various specifications of the relationship. Goodwill was frequently significant but not in the same equations with proprietary technology. The other variables were almost always insignificant at a 90 percent confidence level in the various specifications. These results are not here.

Performance according to company classification

The basic interpretation of the determinants of company performance already has been given. Higher-tech firms tend to demonstrate larger market shares and better perceived performance, and higher-advertising firms do likewise. Attempts to subdivide the sample into only high-ech firms, only consumer products firms, and so on led to mostly ambiguous regression results. Therefore this information is not presented here. The broad conclusion, then, is that consumer products firms showed similar performance to industrial products firms; newer firms showed similar performance to more experienced firms; and wholly-owned firms showed similiar performance compared to their joint-venture counterparts.

Conclusions

Perhaps the main conclusion to be drawn here is that competitive advantages can be seen to influence MNE performance in Latin America; but the only advantages that consistently seem to be significant are proprietary technology and goodwill based on the firm's brand name or company name. This result is broadly consistent with findings of other studies in industrial countries in recent years.

Except for 'Experience,' defined as the length of time of operation in the given country, none of the other competitive advantages showed consistent correlation with performance by the fifty-five firm sample of large, US-based MNEs. This may be surprising, since often other advantages such as economies of scale in production and multinational sourcing and marketing capabilities are discussed as important competitive strengths of MNEs. Perhaps this outcome is due to the fact that the markets are relatively small, so economies of scale are not as important as in most industrial countries. Also, the Latin American countries have imposing geographical barriers between themselves, which make inter-country transportation of products relatively costly — not to mention additional legal barriers such as tariffs, import licensing requirements, and exchange controls. These barriers may largely preclude the use of multinational sourcing and marketing advantages by the firms that possess these capabilities. In any event, the study found only two competitive advantages to be regularly significant in the various countries and groupings of firms.

One may expect this environment to change in the years ahead, as transportation barriers decline and as more European and Japanese MNEs enter Latin America. Just as competition in general has become more international and more intense in the industrial countries in the past two decades, so it probably will develop in Latin America in the next decade or two.

Appendix to Chapter eight. Competitive advantages in the production process: purchasing

Let us consider just the first stage of production — purchasing — to demonstrate the competitive advantages (potentially) involved. Only those advantages that require *international* transactions or communications are discussed.

A firm may gain a competitive advantage in purchasing if it is able to obtain superior *information* about costs and availability of inputs than rival firms. The search cost of obtaining information and/or bids from multiple potential suppliers in more than one country may be unacceptably high for small firms and local firms, so large multinational firms may gain this advantage. The MNE gains an advantage from having its information-collecting network in several countries, by seeking out low-cost and high-quality suppliers in those countries. The firm that can accumulate and assess this information efficiently has an advantage over other firms, including other MNEs, that cannot.

The multi-facility and/or multi-product firm (including the MNE) may gain an advantage in *manpower* inputs by having a small number of purchasing managers/agents who perform the function for many of the firm's facilities or product divisions, thus distributing the cost across multiple uses of this resource. Smaller firms would presumably be less able to utilize fully the purchasing agent's time and abilities. Any firm may gain a manpower advantage in purchasing by hiring a purchaser(s) with superior skills compared to those employed by competitors; though presumably part of this advantage could be lost because of the higher salary paid to the person with superior skills. The multinational firm has access to labor and management at various wage and salary rates in different countries, plus the ability to develop expertise in one country and transfer it to others, plus the ability to separate physically steps in the production process to capitalize on cost and skill differentials in different locations — all of which may create advantages relative to competitors.

The larger firm and/or MNE may be able to economize on *physical assets* needed to operate the purchasing function, if it can concentrate multiple purchasing needs into a single (or a few) optimally utilized facilities, such as offices and warehouses. For example, the multi-product firm gains economies of scope by using a single purchasing staff to buy for all of the lines produced. A multinational firm may be able to locate its purchasing function in a national location(s) where the largest amount of information (and inputs) is available or where facilities' costs are low relative to those in other countries.

Financial capital costs can be minimized in the same way as those of physical capital, if the firm can concentrate the purchasing function in one or a small number of national locations. That is, centralized purchasing requires less financing than separate financing for each of several independent divisions of firms. Second, financial costs can be minimized by purchasing in large quantitites — a possibility available to larger firms in general. The firm may achieve economies of scale in purchasing if it can exercise monopsony power in dealing with suppliers, whether by buying in large quantities or simply by being one of a small number of buyers served by the supplier. The multinational firm may have access to lower-cost funding (for all uses) in foreign and offshore financial markets compared to a local firm. Subsidized funding for specific projects may be offered in some countries by the government, to entice multinational firms to establish facilities (and hence to employ local people and pay local taxes).

Organizing the purchasing function essentially constitutes arranging the various inputs in an optimal fashion. Organizational economies may exist relative to the free market when purchasing information can be passed and stored within the firm more efficiently than in the market; this is considered to be one of the major advantages of the multinational retail seller, Sears Roebuck & Co. (especially since the creation of the Sears World Trade subsidiary). Additional organizational efficiencies may occur because of better coordination among steps of the production process within the (vertically integrated) firm than among unrelated firms in the market, especially when different stages of production are carried out in different countries.

The *risk-bearing* function can be performed by the firm diversifying its purchases among suppliers, thus reducing the dependence on any one of them. Greater diversification can be achieved by choosing multiple suppliers in several countries, to reduce the impact of country risk — certainly, the large firms in the oil industry follow this strategy in sourcing their crude petroleum, as do many other processors of minerals and raw materials. While diversification does not provide insurance *per se*, it does reduce the expected impact of losses from any one supplier. Insurance and/or futures contracts also may be available to the firm, but typically not at advantageous terms that would create a competitive advantage for any particular firm.

Notes

1. See especially Michael Porter *Competitive Advantages*, New York: Free Press, 1985.

2. This list of economic activities is not exhaustive, but it does cover the issues usually raised. Other activities can be analyzed in the same framework.
3. Craig Watson discusses this idea of challenging international competitors in their home markets as a competitive strategy in his article, 'Counter-competition abroad to protect home markets,' *Harvard Business Review*, January–February 1982.
4. Following the international product cycle for any product demonstrates the stages during which proprietary technology, strong advertising and distribution channels, and low-cost production are the key competitive advantages. See, for example, Raymond Vernon, *Storm over the Multinationals*, New York; Basic Books, 1977.
5. These industries *do* show some local production of components, especially in computers where local assembly or production of CRT screens, PC chassis, cables, and other fairly standardized parts is common in many Latin American countries. This production is often done by local firms rather than the MNEs.

Bibliography

Barlow, E.R., and I.T. Wender (1955) *Foreign Investment and Taxation*, Englewood Cliffs, N.J: Prentice-Hall.

Grosse, Robert (1981) 'The theory of foreign direct investment', University of South Carolina *Essays in International Business*, December.

Grosse, Robert (1985) 'Competitive advantages and multinational enterprises', *Discussion Paper # 85-1*, University of Miami International Business & Banking Institute (January).

Hennart, Jean-François (1982) *The Multinational Enterprise*, Ann Arbor, Mich.: University of Michigan Press.

Hout, Thomas, Michael Porter and Eileen Rudden (1982) 'How global companies win out,' *Harvard Business Review*, September–October.

Hymer, Stephen (1960) *The International Activities of National Firms*, Cambridge, Mass.: MIT Press, 1976.

Johansson, Johnny (1983) 'Firm-specific advantages and international marketing strategy', Dalhousie Discussion Paper, 24 (May).

Newman, Howard (1978) 'Strategic groups and the structure-performance relationship,' *Review of Economics and Statistics*, August.

Ouchi, William (1979) *Theory Z*, New York: Basic Books.,

Porter, Michael (1980) *Competitive Strategy*, New York: Free Press.
    (1985) *Competitive Advantages*, New York: Free Press.

Rugman, Alan (1981) *Inside the Multinationals*, London: Croom Helm.

Shapiro, Daniel (1983) 'Entry, exit, and the theory of the multinational corporation', in Charles Kindleberger and David Audretsch (eds) *The Multinational Corporation in the 1980s*, Cambridge, Mass.; MIT Press.

Vernon, Raymond (1966) 'International investment and international trade in the product cycle,' *Quarterly Journal of Economics*.

Williamson, Oliver (1975) *Markets and Hierarchies: Analysis and Antitrust Implications*, New York: Free Press.

175

Chapter nine

MNE structure and strategy in Peru

Introduction

After more than a decade since the effort of Stopford and Wells (1972) to establish a paradigm of the relationship among corporate strategy, organizational structure, and performance in multinational enterprises, no clear consensus exists on either the paradigm itself or on many of the relationships among the factors. That is, although we do understand many of the reasons why firms choose the global strategies and structures that they utilize, still many questions remain about 'optimal' choices of each, and about their relations to performance. Also, many of the relationships have been changing during this past decade.

This chapter explores empirically these relationships in the context of a developing country (Peru) in the 1980s for a group of US-based manufacturing firms. It includes both a description of corporate policies and practices, and an analysis of the impacts of strategies and structures on each other and on performance. Thus our focus is on the 'fit' of a Peruvian affiliate within the overall operations of a multinational enterprise. By focusing on host-country affiliates and their managers, this chapter illuminates the relationships among strategy, structure, and *affiliate* performance. This aspect of the MNE has not often been explored and not at all in the case of Peru.

The model

Corporate strategy and structure clearly are related, and both have some impact on performance. Likewise, government regulation, country characteristics, and industry characteristics are factors which have some impact on company strategy, structure, and performance. The view or model used here hypothesizes that differences in one of these attributes lead to differences in others, for a given firm. The relationships appear as shown in Figure 9.1, with main avenues of influence depicted by arrows.

176

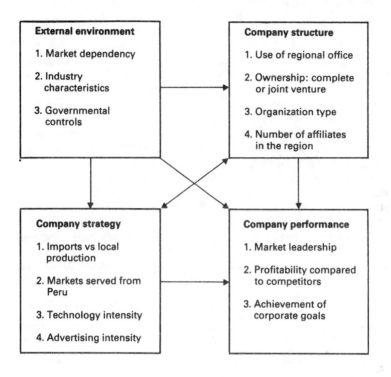

External environment

1. Market dependency

2. Industry characteristics

3. Governmental controls

Company structure

1. Use of regional office

2. Ownership: complete or joint venture

3. Organization type

4. Number of affiliates in the region

Company strategy

1. Imports vs local production

2. Markets served from Peru

3. Technology intensity

4. Advertising intensity

Company performance

1. Market leadership

2. Profitability compared to competitors

3. Achievement of corporate goals

Figure 9.1 Interrelated company and environmental attributes

Figure 9.1 presents a number of relationships between internal and external attributes that make up a firm and its environment. Each of the relationships can be explored (measured) to compare firms along different dimensions. For example, the existence of full ownership as opposed to joint-venture participation can be viewed as a choice that possibly affects performance. Or the level of technology-intensity of the firm may lead to different organizational types (i.e. functional, product, regional, matrix) and/or to different performance. The number of relationships is limited only by the number of attributes measured. Any lines of causality must be asserted, since only correlations can be measured using the data that were collected.

Research design

Data collection for this study was carried out in Lima, Peru, during the summer of 1981. A list of foreign-based firms with affiliates in Peru was obtained from the Peruvian-American Chamber of Commerce. From this list of about 120 firms, eighty-one firms were chosen. The selection criteria were that each firm had to belong to one of six industries (namely chemicals, pharmaceuticals, food processing, electrical appliances, industrial machinery, and office machines), and each had to be a Fortune 500-sized firm. The industries were chosen because each one contains at least eight US-based firms, and because they represent a wide range of technology intensities, advertising intensities, and products subject to varying levels of government controls. The size criterion was used just to insure reasonable homogeneity among firms along that dimension, which was not at issue in the study. Thus the sample includes a group of manufacturing industries and a set of large, primarily US-based multinational firms, that can be viewed as 'traditional' market-seeking foreign investors.

A four-page questionnaire was designed and sent to the general manager of the Peruvian affiliates of the eighty-one firms in July of 1981. Within the eight-week period allowed for responses, twenty-three firms

Table 9.1 Characteristics of fifteen firms in the sample

Company	Industry	Global annual sales ($bn 1981)	Type of affil. in Peru	Year of est. in Peru	Other affiliates in Latin America	Empl. in Peru
1	Chemicals	10.618	Sales	1961	38	48
2	Chemicals	1.828	Sales	1960	7	5
3	Electrical appliances	2.523	Mfg	1906	9	210
4	Food processing	5.819	Mfg	1939	18	115
5	Food processing	15.537	Mfg	1940	47	1112
6	Food processing	4.877	Mfg	1965	11	172
7	Non-electrical machinery	2.239	Mfg	1968	9	40
8	Non-electrical machinery	0.974	Sales	1952	3	52
9	Non-electrical machinery	0.324	Mfg	1967	4	32
10	Non-electrical machinery	0.639	Sales	1963	2	30
11	Pharmaceuticals	3.600	Mfg	1960	31	199
12	Pharmaceuticals	n.a.	Mfg	1927	n.a.	328
13	Pharmaceuticals	12.304	Mfg	1956	4	300
14	Pharmaceuticals	2.969	Mfg	1964	8	40
15	Pharmaceuticals	3.246	Mfg	1962	40	220
Averages	—	5.113	Mfg	1953	16	194

responded, of which fifteen supplied usable answers. Characteristics of the respondent firms are shown in Table 9.1. The analysis below uses the fifteen questionnaire responses, and additional information where available, to explain the marketing strategy and structure of the firms and to generalize about these operations.

An overview of the firms in the sample

While this set of firms was not chosen as 'representative' of a larger population, it is expected that their characteristics are indicative of other manufacturing MNEs that operate in Peru. The present section outlines some of the broad similarities among firms, and then subsequent sections explore inter-firm differences.

As noted in Table 9.1, the sample firms had average world-wide sales of $5.113 billion in 1981. Two-thirds of them engage in some manufacturing in Peru, while all of them serve Peru as their primary market. (Half of the firms do some exporting to other Andean countries; none sell elsewhere.) With only one exception, these firms import at least some of the products that are sold in Peru; and the imports come primarily from the United States. Thus the sample is composed of what may be called 'traditional' manufacturing MNEs, that have entered the host country to serve the local market through some combination of local production and imports from other affiliates.

Over half of the sample firms entered Peru in the 1960s, and all had established a Peruvian affiliate by 1968. Most began their local operations by creating a new affiliate; only one entered through acquisition of an existing firm, and that acquisition was another MNE. Because all of the firms had begun operations before the formation of the Andean Pact, they avoided the forced fade-out of ownership required under the 1971 Foreign Investment Code. In fact, all of the affiliates are at least majority-owned by the parent, and about half are wholly-owned. Thus the parent firms are able to exercise control over all of these Peruvian affiliates, because of their ownership positions.

As is often the case in 'traditional' manufacturing investment, the affiliates are viewed as profit centers by the parents. Most of the affiliates maintain substantial control over marketing, personnel, service, and even financial decisions — with a dollar limit on transactions that may be carried out without home office approval. Major capital budgeting and intra-firm pricing decisions generally are made by the home office (though, of course, with substantial input from the affiliate). Table 9.2 depicts the assignment of some basic decisions to the home office, the affiliate, and in some cases to a Latin American regional office.

Note the paradox of the two points mentioned above; though its hands are tied because major financial decisions are assigned mainly to the home

Table 9.2 Decision-making responsibilities in fifteen MNEs (with respect to activities of their Peruvian affiliates)

Decision	Mainly home office 1	Half & half (home & affil.) 2	Mainly regional office 3	Mainly Peru affiliate 4
MARKETING				
1. Choice of products to sell			*3.25	
2. Choice of markets to serve from Peruvian affiliate			*2.875	
3. Pricing strategy for sales by Peruvian affiliates			*3.50	
4. Size of Peruvian affiliate advertising budget			*3.067	
5. Choice of distribution channels in Peru				*3.688
6. Choice of suppliers to Peruvian affiliate			*3.188	
FINANCE				
7. Allocation of funds to Peruvian affiliate		*2.0		
8. Immediate profit responsibility for Peru affiliate				*3.75
9. New plant/office construction/ purchase in Peru		*2.25		
10. Pricing of intra-firm sales to Peru affiliate	*1.571			
OTHER				
11. Size of Peruvian affiliate R & D budget			*2.909	
12. Provision of legal services in Peru			*3.50	
13. Provision of technical support service to customers			*3.214	

Source: Questionnaire responses from fifteen affiliates in Peru, 1981.

office, the affiliate still retains profit responsibility. Also note that, even though proprietary technology is a competitive factor for many of the firms in the sample, responsibility for R & D and technical support for the affiliate is decentralized to the affiliate or to a regional office. Additional characteristics and policies of the firms are presented below through inter-company contrasts along several dimensions.

Major characteristics of the firms by industry

The most striking and easiest-to-understand differences between firms reflect inter-industry variations in attitudes and in policies. Let us consider each major industrial category in turn. The *chemicals* industry in Peru includes twelve MNE affiliates, of which two responded to the questionnaire. Both of these firms operate only sales offices in Peru, with no manufacturing. Both sell industrial and agricultural chemicals, many of which constitute 'new products' in the product life cycle. Such products are the result of substantial R & D efforts by these firms, which are considered 'technology-intensive,' based on the US average R & D spending by manufacturing firms.

The resins and agricultural chemicals produced by these firms require fairly large-scale production to achieve least-cost output — and the Peruvian market often is too small to absorb that level of output. Hence, imports from the United States are used to supply this market, rather than local production. One of the two firms exports some of its sales to Bolivia, but both generally aim at the Peruvian market as the main target. The wholly-owned firm is the market leader in Peru for its most important products; while the majority-owned firm is not, ranking only among the twenty leading firms for its products. The former is controlled tightly from a Miami regional headquarters, while the latter has substantial autonomy, except for undertaking major capital projects and provision of technical service to customers, which are controlled from the home office.

In contrast, the *pharmaceuticals* firms in Peru include eighteen foreign MNEs, five of which responded to the questionnaire. All of the respondents are manufacturing most of their drugs in Peru, and three of them import less than 10 percent of what they sell. All of these firms are considered to be 'technology-intensive,' with large global expenditures on R & D and a large number of new products sold. All but one of them rely on proprietary drugs for the main part of their Peruvian sales.

While the pharmaceuticals industry is very technology-intensive, depending on R & D to generate new drugs on a continuing basis, the processing done in Peru is not. That is, the firms concentrate their research in the US and/or in Western Europe, producing basic chemical entities there; then the final pharmaceutical pills, creams, tablets, and so on, are assembled (formulated) in Peruvian plants. Since the formulation stage does not require extensive R & D, or large capital outlays, drug companies have entered the Peruvian market, and other relatively small Latin American markets, with formulation plants. Interestingly, the bulk of value added from the five firms' sales comes from the local plant; all but one firm import less than a quarter of their Peruvian sales value.

The Peruvian market is adequate to consume the output of these plants, though most of the firms also export some drugs to Bolivia or Chile. Pharmaceuticals firms tend to produce locally for local sales in Latin America, and they transship very little between affiliates. (Most of them just export the chemical entities from US plants to the Latin American affiliates and formulate the drugs locally in each one.)

All but one of the drug companies share some (minority) ownership with local investors in Peru. All of the joint ventures have substantial autonomy in decision-making in the local affiliate, with the exception of decisions on transfer pricing, that are made mainly by the parent firm. All five of the firms are very 'advertising-intensive,' spending more on advertising (as a percentage of sales) than all of other firms in the sample (with one exception). Despite their promotional expenditures, two of the five firms are not market leaders in Peru, while the other three are. The latter three firms also are the least integrated multinationally in production, with less than 10 percent of their local sales being imported.

The *food-processing* industry is composed of nine foreign MNEs as well as numerous local firms in Peru. Three of the nine MNE affiliates responded to the questionnaire. Similar to the pharmaceuticals manufacturers, these three firms produce almost all of their local output in Peru, and they all import less than 10 percent of their sales. Local production is feasible because the basic food inputs are grown locally, and the fairly standardized production processes are not overly capital-requiring.

Food processing generally is a maturing or standardized industry in the product (or industry) life cycle, and these three firms fit this mould quite well. All of them rank well below the US average in technology intensity. Simultaneously, they all rank well above the US average in advertising intensity. Typically, standardized products face strong price competition, they use well-known technology in production, and they are differentiated through advertising by the producers. Second only to the pharmaceuticals firms in the present sample, the food processors average over 4 percent of global sales in advertising expenditures. Clearly, product differentiation is an important competitive tool in this industry.

Two of the three food-processing affiliates are wholly-owned by the US parent, and the third is majority-owned. The former two serve only the Peruvian market with their sales, while the latter also exports to other Andean countries. The wholly-owned affiliates also are market leaders with their main products, while the joint-venture is not. Despite these differences, all three firms retain substantial autonomy in decision-making. (The joint venture reported a higher degree of home office control over allocation of funds to the affiliate, but similar conditions in all other decision categories are shown in Table 3.2.)

The remaining five firms can be grouped as *miscellaneous manu-*

facturing, though more precisely they produce electrical appliances, industrial machinery, and non-electronic instruments. Thirty-four MNEs constitute the foreign participation in these industries, and five of them responded with usable questionnaires. Since these firms do not form a very homogenous set, no generalizations are attempted here. Instead, the firms are grouped with others in the sample to generalize about strategy and structure below.

Hypothesis tests on company and environment relationships

The impact of market dependency

Consider first the importance of the firm's external environment as an influence on the three other classes of attributes. Possibly the most important division to be drawn is between very market-dependent firms and more footloose and/or producer-goods firms. That is, both the strategy employed by a MNE in Peru and its organizational structure (including a Peruvian affiliate) are substantially determined by the market dependence of the firm. Market dependence, in turn, is the importance of country (market)-specific conditions, especially customer preferences, to company decisions. Food-processing firms, for example, must tailor their products literally to consumer tastes, which often differ across cultures. Chemicals manufacturers, on the other hand, sell basically the same products world-wide, often to industrial clients rather than individual consumers. Pharmaceuticals producers often are constrained to manufacture locally in order to meet local health standards, even though their drugs generally do not vary across countries. Also, governments in Latin America have raised very high tariffs and non-tariff barriers on trade in pharmaceutical products, thus encouraging local manufacture. And, the pharmaceutical firms are very 'advertising' intensive,' spending relatively large amounts of money on promotion. Finally, the miscellaneous manufacturers are primarily footloose, selling the same products world-wide.

The market-dependent firms in food processing and pharmaceuticals appear to have chosen (or been forced by government pressures) to operate very autonomously in Peru, importing very little of their production inputs. A Mann-Whitney U-text comparing imports of these firms to imports of the other firms in the sample yields a test statistic $U=O$, compared to a critical value of $U=13$.[1] Thus the null hypothesis of no difference in importing propensities of the two groups of Peruvian affiliates is rejected.

The market-dependent firms also demonstrate much greater integration into the Peruvian economy, with somewhat more local ownership, substantially more local manufacturing, and substantially more local employment. Table 9.3 shows results of several Mann-Whitney U-tests, used to discern differences in the attributes of the two groups of firms.

Table 9.3 Hypothesis tests on market-dependent versus other firms

Test	Mann-Whitney U-value	Critical U (α = 0.05)	Accept/ reject
1. Difference in imports as a % of sales	0	13	Reject
2. Difference in local ownership	22.5	13	Accept
3. Difference in sales vs manufacturing	12	13	Reject
4. Difference in local employment	6.5	13	Reject
5. Difference in advertising intensity	6	11	Reject
6. Difference in number of affiliates in Latin America	10.5	11	Reject
7. Difference in market leadership	25.5	13	Accept

Sources Questionnaire responses; Moody's *Guide to Industrial Companies: Advertising Age*, annual survey of advertising intensity.

Notice that the test on a difference in ownership policy yielded insignificant results, though numerically more market-dependent firms operate as joint-ventures than do other firms. Also, note that, despite the multiple strategic differences, the two groups of firms show no difference in performance (i.e. in market leadership).

The market-dependent firms in Peru also demonstrate a larger global propensity to use advertising, though their questionnaire responses only showed a higher commitment by pharmaceuticals manufacturers. (Still, the hypothesis of equal advertising-intensity between groups of firms is rejected with either measure.)

Finally, the market-dependent firms have a broader affiliate network in Latin America than do other firms in the sample – presumably because they face similar constraints in other national markets as in Peru.

No other tests were performed using external environmental attributes, since inter-industry differences already have been described, and since governmental controls were not measured directly.

Company strategies

Among the many strategic issues that could be considered, two focal areas were selected, both of which relate importantly to multinational business. First, the company's choice of production locations was considered in terms of (1) the percentage of sales in Peru generated from local production versus the percentage imported from production facilities elsewhere (generally in the United States), and (2) the use of Peruvian output for local sales versus export to other (Andean) countries. Thus the firms' international sourcing and distributing strategies were analyzed in this elementary way. Second, the company's allocation of resources for, first, promotion, and for, secondly, R & D, offers insight into its exploitation of competitive advantages, such as a reputable brand name or proprietary technology. Thus expenditures of the firm on these two activities gives some idea about its competitive strengths (and to some extent about the industry's characteristics as well).

Looking first at sourcing of production, it was found that all pharmaceuticals and food-processing firms import less than half of their Peruvian output, and all of the other firms import more than half (as tested in Table 9.3, test 1). This leads to no new conclusions, though the dichotomy is striking — especially if we note that virtually all of the others import more than 75 percent of theirs.

Comparing the choice of markets to serve from the Peruvian affiliate *did* generate new results, since all four industries contain firms which sell only in Peru, and other firms that export some products. Indeed, it was quite interesting to note that the firms which do *not* export from Peru are primarily the wholly-owned affiliates. They also tend to be market leaders and generally were more recently established in Peru than the other firms, though both of these measures were not significant. Table 9.4 presents selected comparisons of firms that use these two alternative strategies on structural and performance measures.

Table 9.4 Hypothesis tests on affiliates that export from Peru, versus others

Test	Mann-Whitney U-value	Critical U (α=0.05)	Accept/ reject
1. Difference in percent of local ownership	10.5	12	Reject
2. Difference in market leadership	19.5	12	Accept
3. Difference in year of establishment	18	12	Accept
4. Difference in number of affiliates in Latin America	21.5	12	Accept

Source: Same as Table 9.3.

It should not be surprising that the more recently arrived affiliates in Peru have concentrated on the domestic market, while their more-experienced counterparts have developed some export markets. It is perhaps more striking that the former group is predominantly 100 percent owned by the parent, though pressures for local ownership were mounting in the 1960s when most of them were established in Peru. Also, it is rather surprising to see that in performance, almost all of the non-exporting (more recently arrived) affiliates are market leaders. More evidence is needed to shed light on these anomalous findings, since we would be most interested in knowing whether the choice of market(s) to serve or the choice of 100 per cent ownership has led to market leadership in Peru (if either did).

The other two measures of company strategy generated ambiguous results. The advertising intensity of several firms was unavailable for the whole MNE, while only the pharmaceutical firms responded that they spent more than 10 percent of their Peruvian budgets on advertising. As far as technology intensity is concerned, no significant differences were found between relatively 'high-tech' firms and relatively low-tech ones for the attributes measured. Though not significant, the high-tech firms have a larger percentage of joint ventures than the low-tech firms, contrary to what we would expect, that is that firms would protect their proprietary techology by keeping full ownership of affiliates.

Company structures

The main question to be investigated in the area of organizational structure was the choice between full (100 percent) ownership of the affiliate and some level of joint-venture ownership. This issue is key because: (1) US firms have traditionally sought 100 percent ownership, but in the past two decades many less-developed countries have forced them into joint ventures; (2) the Andean Pact (in which Peru is a member) in particular requires partial local ownership for affiliates established since 1971, and encourages it for previously-existing affiliates, (3) technology-intensive firms would presumably prefer strongly to maintain full ownership; and (4) joint ventures would presumably have different decision-making freedom relative to wholly-owned affiliates. The performance and other characteristics of international joint ventures has been a subject of intense interest to MNE managers and host-country regulators for almost two decades, as the rules of the game for MNEs develop.[2]

Selected results of tests on strategic, structural, and performance differences between wholly-owned affiliates and joint ventures are shown in Table 9.5. The most striking result is that wholly-owned affiliates tend to be market leaders, whereas joint-ventures more frequently are not. In some way the existence of total control over the affiliate appears to enable firms to take leading positions in their Peruvian markets. It is noteworthy also that the 100 percent owned affiliates tend to concentrate on serving the Peruvian market (though this inference was not quite significant at the ninety-five percent confidence level used in the test.

Table 9.5 Hypothesis tests on wholly-owned affiliates versus joint ventures

Test	Mann Whitney U-value	Critical U (α=0.05)	Accept/ reject
1. Difference in market leadership	12	13	Reject
2. Difference in local employment	8.5	13	Reject
3. Difference in markets served	13.5	13	Accept
4. Difference in technology intensity	8	7	Accept
5. Difference in number of affiliates in Latin America	12.5	11	Accept
6. Difference in imports as a percent of sales	26.5	13	Accept

Sources: See Table 9.3

In terms of strategies, there were no measures that distinguished between wholly-owned and partially-owned affiliates. While the average technology-intensity of wholly-owned affiliates was lower, it was not significantly so. While there were more sales offices and less local manufacture among the wholly-owned affiliates, the difference was quite insignificant statistically. In all, no measurable strategic differences appeared between these two groups of firms.

Even in terms of other aspects of organizational structure, the level of ownership did not differ significantly. Both wholly-owned affiliates and joint ventures have about the same distribution of organizational forms (i.e. product, functional, regional, or matrix) and frequency of regional offices (about 80 percent use a Latin American regional office). Conversely, firms that use regional offices did not differ significantly from others in any of the measures used, nor did firms employing an international division (plus domestic product or functional divisions) differ from those organized otherwise. The locus of decision-making on all of the questions asked in Table 9.1 was virtually identical between groups, as well.

Company performance

The final attribute investigated here was performance, and only one proxy was obtained, that is market leadership. Dividing the firms along this dimension yielded *no* significant differences between groups for any of the attributes measured. This result is rather surprising, since both '100 percent ownership of the affiliate' and 'serving only the Peruvian market' correlated very highly with market leadership. However, when the firms were grouped strictly on the basis of market leadership, the relationships were too weak to yield significant inter-group differences.

Conclusions

The results of this analysis demonstrate some interesting and potentially useful relationships among the attributes measured. Unfortunately, when looking at a relatively small developing country, with a MNE group of under 200 firms — and a responding group of only twenty-three firms — the conclusions must be very tentative.

The correlations obtained in the preceding section support the three pictures shown in Figure 9.2 of causes and effects in Peruvian

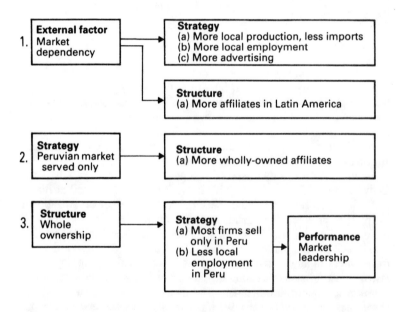

Figure 9.2 Relationships among strategy, structure, performance and environment.

affiliates of foreign MNEs.

Clearly, the single factor that led to the greatest difference in responses by MNE affiliates was the degree of dependency on the Peruvian market. The more market-bound the affiliate, the more integrated it is into the Peruvian economy. This one external factor had more impact on company attributes than any of the internal characteristics that were measured.

The strategy of selling only in Peru (without exporting) and the structure of 100 percent parent-company ownership are very highly correlated in both directions. Perhaps this means that joint-venture partners tend to force Peruvian affiliates to enter export markets, whereas parents of wholly-owned affiliates tend to ensure concentration on the domestic market. This speculation is consistent with the result that the wholly-owned affiliates have tended to succeed as market leaders in Peru, where they are concentrating their efforts.

Ultimately, we would like some additional measures of affiliate performance and some additional responses to strengthen the findings in the analysis.

Notes

Acknowledgements: I would like to thank the Fulbright Commission of Peru and the Universidad del Pacifico for financial support, without which this project could not have been undertaken.

1. The Mann-Whitney U-test is a non-parametric test of the difference in means between two samples. This test does not consider the distributions of the two samples. A test value *less* than the critical value lead to rejection of the null hypothesis, which is that the two sample means are equal.
2. See, for example, W. Friedman and D. Beguin, *Joint International Business Ventures in Developing Countries*, New York: Columbia University Press. 1971.

Bibliography

Aylmer, R. (1970) 'Who makes marketing decisions in the multinational firms?' *Journal of Marketing*, October.

Business International Corporation (1981) *Organizational Structure in Multinational Corporations*, New York: Business International Corporation.

Caves, R. (1980) 'Industrial organization, corporate strategy, and structure', *Journal of Economic Literature*, March.

Chandler, A. (1962) *Strategy and Structure*, Cambridge, Mass.: MIT Press.

Davis, S. (1976) 'Trends in the organization of international business,' *Columbia Journal of World Business*, summer.

Dominguez, L. (1983) 'Marketing structure and strategy in Venezuela', unpublished paper, Miami.

Friedman, W., and D. Beguin (1971) *Joint International Business Ventures*

Corporate strategies

in *Developing Countries*, New York: Columbia University Press.

Goehl, D. (1978) *Decision Making in Multinational Corporations*, Ann Arbor: UMI Research Press.

Grosse, R. (1981) 'Regional offices in multinational enterprises: the Latin American case', *Management International Review*, 2.

Lawrence, P., and J. Lorsch (1967) 'Differentiation and integration in complex organizations', *Administrative Science Quarterly*, June.

LeCraw, D.(1983) 'Performance of transnational corporations in less developed countries', *Journal of International Business Studies*, spring/summer.

Newfarmer, R., and Mueller, W. (1975) 'Multinational companies in Brazil and Mexico: structural sources of economic and non-economic power,' Report to the Subcommittee on Multinational Companies of the Committee on Foreign Relations, US Senate, Washington, DC: USGPO.

Stopford, J., and Wells, L. (1972) *Managing the Multinational Enterprise*, New York: Basic Books, 1972.

Wiechmann, U. (1976) *Marketing Management in the Multinational Firm: The Consumer Packaged Goods Industry*, New York: Praeger.

190

Chapter ten

Financial transfers in the MNE: the Latin American case

Introduction

This chapter examines a variety of types of intra-company, international transfers of funds. Its goal is to analyze key characteristics — especially the costs — of the money-moving vehicles that are available to firms with affiliates in more than one country, that is multinational enterprises (MNEs). The entire analysis focuses on the funds-transfer process; it is assumed that the firm already has decided on the optimal allocation of funds for each affiliate during the relevant time period. Also, the empirical evidence comes from firms with affiliates in Latin America, a group of countries plagued with foreign-debt problems and character-ized by tight restrictions on funds transfers by multinational firms.

Two of the key dimensions that distinguish international from domestic finance are the needs to deal with more than one national legal environ-ment and to operate in more than one currency. In this broader context, the issue of intra-company funds transfer takes on added importance, because the transfers can be used to reduce currency risk and to arbitrage national legal (i.e. tax) conditions. In Latin America, firms are severely restricted as to cross-border transfers of funds, so the issue plays a large role in corporate finance for MNEs operating in countries of that region. While exchange risk management must play an important part in company decisions as to when to move funds between countries, it does not necessarily affect the choice of transfer method — and exchange risk is not treated in any detail in the chapter.

The potential methods for moving funds internationally among affiliates of the MNE are infinite, dependent only on the creativeness of the contracts or other arrangements utilized. In reality, of course, legal constraints limit many of the potential transfers, but still leave the door open to others. In the Latin American region, the problem may best be viewed as a choice among methods of moving *value*, rather than just funds, from affiliates to the parent corporation. That is, since foreign exchange such as dollars is very scarce in most countries of the region, the multinational firm may want to consider transfers of products and services as well as money. If the value can be transferred from one country to another in the form of products, and then converted into currency, the same result is achieved — albeit in a more protracted and difficult manner. Due to the continuing crisis of external debt in the region during the 1980s, this emphasis on value rather than simply funds is an important shift of focus.

In this paper, seven general methods of transferring value are considered:

1. Loans
2. Dividends
3. Transfer pricing on intra-firm product shipments
4. Fees or royalties, that is service charges, for use of intangible property such as technology
5. Parallel loans
6. Counter-purchases
7. Swaps of assets or liabilities between affiliates of multinational firms

These seven methods of moving value from an affiliate or parent firm in one country to another related entity in another country are depicted in Figure 10.1. They do not exhaust all of the possibilities, but they provide a framework for thinking about such transfers.

In order to move value from a Latin American affiliate to the US parent firm, the affiliate may make a loan to the parent, pay a dividend to the parent (i.e. remit profits), pay for shipments of goods from the parent, pay royalties or other fees for 'services' provided by the parent, arrange a parallel loan through a financial institution, or arrange some kind of swap, as described below. If a loan previously were made from the parent to the Latin American affiliate, then additional funds transfers would occur through interest payments and principal repayment. Each of these funds flows is depicted in Figure 10.1; others that managers devise may be added in the same framework.

Table 10.1 compares these same transfer methods according to their ties to other transfers in the firm, their timing, their tax implications, and the relevant decision variable. Notice that a loan contract, which spells out interest and principal repayment terms, is much more restrictive of an arrangement than a dividend payment, which can be made at any time and in any amount for which earnings are available (assuming the funds can be converted into home-country currency). Also notice that the swap agreement requires the MNE to find a partner firm with which to carry out the exchange; the other transfer methods utilize banks or other readily-accessible transfer mechanisms.

Given even this small set of options, how can the financial manager

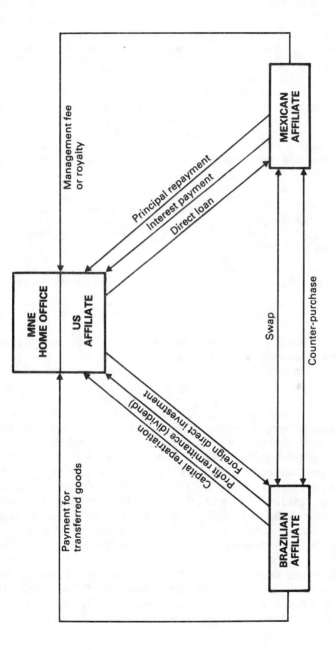

Figure 10.1 Some methods of moving funds in the MNE

Table 10. 1 A comparison of intra-firm, international financial transfers (funds flows from host to home country)

Type of transfer	*Ties to other intra-firm flows*	*Timing*	*Relevant taxes (d=domestic: f=foreign)*	*Decision variable*
1. Loan to parent	Future interest plus principal repayment	Open	d.f. corporate income: d. int withholding	The loan contract
2. Dividend paid to parent	Prior investment from parent	Open	d.f. corporate income; f. div withholding	Payout of subsidiary's after-tax profits
3. Transfer prices for shipment to affiliate	Goods move in opposite direction	Open	d.f. corporate income; tariffs	Sales/purchase contract
4. Fees and royalties	Services move in opposite direction	Fixed by tech. or mgt. contract	d.f. corporate income; f. royalty withholding	Technology or management contract
5. Parallel loan	Bank contract in each country	Open	d.f. corporate income	The loan contract
6. Counterpurchase	Goods move in each direction	Open	d.f. corporate income	The barter agreement
7. Asset or liability swap.	Depends on what is swapped	Open	Depends on the swap arrangement	The swap agreement

make a choice? The following sections explore both theoretical and empirical answers to this question.

Hypotheses

In the literature a large number of studies have considered the issue of international transfer pricing, from the perspective of the firm trying to maximize after-tax profits (e.g. Horst 1971; Copithorne 1971; Nieckels 1976; Merville and Petty 1978; Batra and Hadar 1979; Rugman and Eden 1985; and many others) and from the perspective of governments trying to maximize social welfare (e.g. Copithorne 1971; Kopits 1976). Others have presented models of maximizing behavior for MNE affiliates' dividend policy (Obersteiner 1973), royalties (Kopits 1976) and other kinds of transfers. Rutenberg's (1970) classic presentation of all of these transfer methods allows comparisons among them and demonstrates their complementarity.

The present analysis simply discusses the conditions under which each of the transfer methods would offer optimal results for the firm. The first hypothesis is that a set of conditions based on cost parameters (especially regulatory conditions) and other constraints can be specified to compare each of the transfer methods discussed here. From these conditions, optimal transfers can be derived. Using the framework of Grosse (1985), the tax costs of each transfer method are compared to show 'optimal' transfers under highly restrictive assumptions. A sub-hypothesis is that, since regulatory conditions differ substantially from country to country, the optimal transfer method is likely to differ depending on the country.

Despite these constraints, it still is expected that companies will try to take maximum advantage of the transfer opportunities available to them. Thus a second hypothesis is that firms will attempt to utilize the cost-minimizing strategies for moving value from their Latin American affiliates. While very little published information exists to test this hypothesis, some data aggregated across firms are available from the US Department of Commerce, and additional evidence was collected from a survey of financial managers in a set of multinational firms.

Comparison of methods

Loan from affiliate to parent[1]

A loan will generate no new tax costs to the firm (beyond taxes previously paid in the host country on earnings, which generated funds to make the loan). Subsequent interest payments from the parent firm, however, will reduce income in the home country and add to income in the host country. Note that by charging a very high interest rate and/or bunching principal repayments early in the loan's life, the firm can generate a substantial flow of funds *from* the parent to the affiliate. If the original loan were made from parent to affiliate, then this strategy would be similar to investing and repatriating a large percentage of earnings through dividends and decapitalization.

Note also that the form of the loan may vary in intra-firm agreements. For example, an implicit loan occurs when the affiliate lags or leads payment for a service or product provided by the parent. The implicit interest rate then must be determined from the firm's opportunity cost of capital.[2] This form of loan avoids all tax implications, unless the payment timing falls within a different accounting period.

Dividends

A loan contract specifies the exact timing and value of the funds transfers involved. Dividends, on the other hand, can be chosen in any time period with no necessary constraint from the initial investment.[3] However, dividends are taxed first as earnings in the host country, then often with an additional withholding tax, and finally in the home country when remitted. Because a loan typically faces only a tax on the interest earnings of the lending affiliate (and a tax deduction for the interest-paying affiliate), the loan generally is a less costly method of transferring funds. Host governments understand this point, and frequently they will not allow a loan from affiliate to foreign parent, choosing to construe the loan as a dividend instead.

Transfer prices on goods

The third method of financial transfer differs significantly from the first two in that a flow of goods constitutes the opposite side from the money flow. In this case the price can be determined more or less at the discretion of the firm, depending on legal constraints. For example, governments often disallow transfer pricing on goods at prices deemed to be 'dumping'[4] at less than fair value; and they also try to determine if companies overcharge their affiliates to move funds to the shipping affiliate. Not only are prices fairly discretionary to the firm, but also timing of the payment (i.e. leading or lagging) may be arranged to achieve a variety of goals, such as avoiding exchange risk or making capital available sooner or later than would a pure market transaction.

The tax implications of transfer pricing depend on whether a high or low price is used. The higher the transfer price, the greater the income shown in the shipping and the less shown in the receiving affiliate.

Royalties and fees

Royalties or fees may be charged within a firm as payment for services by one affiliate to another. Typically, these charges are made by the parent firm to its subsidiaries for the use of proprietary technology or management know-how. Royalties or fees thus are tied to an intra-firm service flow, but the service is intangible, and the value may be assessed rather arbitrarily by the firm.

These charges for the provision of intangibles by the parent firm lead to tax-deductible expense for the paying affiliate and taxable fee income for the parent. Depending on the tax treatment in each country, the company may gain or lose financially — though the funds transfer will be carried out regardless. This result is quite similar to the tax effect of interest payments between parent and affiliate under a loan contract.

A potential choice between royalties/fees and transfer prices on goods is only relevant in the sense that the potential tax gain from using the royalty might be higher than the gain from using a high transfer price. This problem is specious, because these funds-transfer methods are complementary, not mutually exclusive. The same holds true for any comparison between transfer prices on goods and either loans or dividends.[5]

Parallel loan (swap loan)

The final three methods of value transfer actually are means to avoid transfering funds internationally. The term 'swap' is used to define any exchange of assets or liabilities between parent and affiliate, typically through a third party such as a bank or other MNE, excluding simple exchange of goods for money (i.e. transfer pricing). For example, the *parallel loan* is a combination of a loan extended by a MNE affiliate to a bank in one country with a loan by the bank's affiliate (or parent) in another country to the MNE's affiliate in that country. Figure 10.2 depicts this transaction. Here, no funds move across national borders, but the affiliate effectively shifts the funds to its parent via a multinational bank. A complicating factor is that the loans must be repaid, at which point the transfer is undone. The tax impact of this transfer is similar to the loan discussed previously, with the affiliate earning taxable interest on its loan, and the parent firm paying tax-deductible interest on its loan. The foreign exchange impact of the swap may be very significant, because in this case the bank is accepting funds in local currency and lending them out in home-country currency — and the bank will charge for this service. This swap similarly may be carried out through affiliates of two MNEs, one of which needs cash in the host country and the other of which has excess cash there. The first firm borrows from the second in the host country, and their parents simultaneously lend between themselves in the opposite direction. Again, the valuation of the home currency versus host currency loans may be difficult, and the swap is undone when the loans are repaid.

Counter-purchase (product swap)

A second form of swap between two MNEs allows for a permanent exchange of assets. In this case one affiliate purchases some product or service from the other in the host country, using up cash and reducing the selling firm's inventory. At the same time the parent firms arrange to sell some product or service between themselves, such that the first company receives income and the second uses cash to buy needed inputs. The double arrangement could be viewed as a form of counter-trade, with the net effect being that each company has sold output in one country and purchased the other's output in the other country. There are no explicit tax implications for this transaction, nor are there limitations set by governments. However, a substantial cost of using this type of transaction is the search cost of finding a partner firm to deal with, and negotiating terms. Figure 2b depicts this swap.

Figure 10.2 Swap transactions
(a) Parallel loan

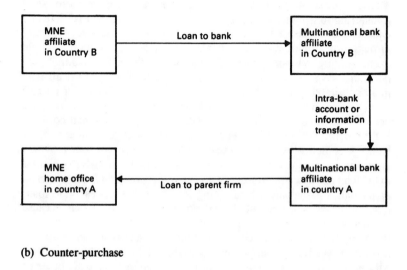

(b) Counter-purchase

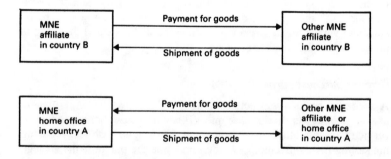

Asset or liability swap

In general terms, any exchange of assets or liabilities between firms or affiliates of the same firm can be called a swap. Beyond loans and products, other assets such as securities, accounts receivable and payable, and other balance sheet items can conceivably be exchanged. Just as sale of a capital item and leaseback allows US companies to shift tax liabilities, so too can asset exchanges be used, in principle, among parent and affiliates in different countries. Perhaps the simplest exchange between two MNE affiliates is the payment for products shipped between them with other products instead of money. If the Latin American affiliate cannot obtain foreign currency to pay the intra-company shipment, a 'barter' arrangement can be used to get value out in the form of other products. The difficulty, of course, is in finding some product(s) to pay with, such that acceptable value really is received by the other affiliate or parent firm. (Colombia, since 1984, has required importing firms to create exports of equal value in order to obtain an import license from the government; so this kind of swap would be very useful for an affiliate there.)

Limitations placed on financial transfers by Latin American governments

Using the perspective in the previous section, a set of optimal transfer methods could be derived for any country, given the relevant tax parameters and other regulatory constraints. To give an idea of the limitations involved, a selection of key regulatory policies for eight Latin American countries is presented in Table 10.2 below.

This table examines only a few of the restrictions placed on international financial transactions by Latin American governments at present. All of these countries restrict access to foreign exchange, most by using a preferential exchange rate for official transactions and a much less favorable one for private-sector dealings. Looking at the main transfer methods discussed in this chapter, we see that most countries limit remittance of dividends, interest and royalties/fees through taxes or quantity limits. Swap agreements generally are not greatly restricted. Leading payments for intra-company shipments of goods generally is not permitted, and lagging payments is encouraged or required.

Using the optimizing measures from Grosse (1985), transfers can be ranked for each country on the basis of tax costs. While every company may face a different tax situation in each country, due to complex transactions and rules, at least the transfer methods can be ranked for a hypothetical case. Assume that the company faces the highest marginal income tax rate in each country, and that other constraints (e.g. rules disallowing one or more transfer methods) are not binding. Then, according to published tax rates, Table 10.3 presents rankings of *taxable* transfer methods such that costs are minimized.

Dividends generally rank lowest on the scale because they are paid out of after-tax income. Royalties and loans almost always are preferable because they are paid out of pre-tax income. In this particular set of countries, for this hypothetical firm, loans are optimal (among these three

Corporate strategies

Table 10.2 Selected financial policies in Latin American countries

Country	Limit on profit remittance	Limit on royalties and fees	Limit on intra-firm interest	Limit on capital repatriation	Limit on leads & lags	Limit on access to foreign exchange
Argentina	Tax on dividends greater than 12% of of capital	18% tax	None	Not allowed for three years	5% lead payment maximum	Dollars bought via foreign exchange certificate
Brazil	Same as above	Maximum 5% of sales	LIBOR + 2.25% max.	n.a.	n.a.	Bureaucratic delays*
Chile	None	Maximum of 5% of sales	None	Not allowed for three years	90–120 day lead or lag allowed	Same as above
Colombia	Maximum dividend 20% of capital	Prohibited	Prime or LIBOR + 2%	None	Not permitted beyond 180 days	Dollars bought via foreign exchange certificate
Ecuador	Same as above	Prohibited	n.a.	Not allowed for three years	n.a.	2-tiered foreign exchange market
Mexico	Tax on dividends	Up to 5–6% of sales	None	None	None	2-tiered foreign exchange market
Peru	Maximum dividend 20% of capital	Prohibited to parent firm	2% above LIBOR	n.a.	n.a.	Bureaucratic delays
Venezuela	Same as above	Same as above	2% over LIBOR or ½% over prime	None	None	3-tiered foreign exchange market

Sources: Business International Corporation, *Financing Foreign Operations*, New York: BIC 1984
* Bureaucratic delays include various forms of inaction by central banks to avoid transferring dollars to firms or individuals. These delays are indefinite in many countries in 1984, due to the debt crisis.
n.a.: not available

Table 10.3 Optimal transfer methods by country (assuming that the firm is in the highest marginal tax bracket)

Transfer Method	Argentina	Brazil	Chile	Colombia	Ecuador	Mexico	Peru	Venezuela
1 = best	R	L	R	L	L	L	R	R
2 = 2nd best	L	R	D	R	R	R	L	L
3 = 3rd best	D	D	L	D	D	D	D	D

Sources of tax rates: Business International Corporation, *Investing, Licensing and Trading Conditions*, New York: BIC, 1984 country updates.
R = royalty; L = loan; D = dividend.

200

transfer methods) in half of the cases, and royalties are optimal in the others. The *untaxed* transfer methods such as counter-purchases and parallel loans offer even better results in some cases.

By far the most important consideration in financial transfers from Latin America during the past four years has been access to dollars (or other hard currency). Due to the debt crisis (see Cline 1983), most of the forms of transfer discussed above have been limited severely. Under these conditions, the optimal strategy may be one that eliminates the use of hard currency (e.g. some form of swap) rather than one which minimizes transfer costs.

Some empirical evidence on transfers in Latin America

Two sources of data were utilized to investigate companies' actual practices in transferring funds from affiliates in Latin America to their parents. First, aggregate data collected by the US Department of Commerce is presented, showing a comparison among dividends, interest payments, and royalties and fees paid by Latin American affiliates to their U.S. corporate parents. No aggregate measures of transfer pricing or swap transactions are available from government sources. Second, results of a survey of corporate treasurers for Latin America are presented.

US Department of Commerce Data

Table 10.4 presents data compiled by the US Department of Commerce on remittances from US firms' affiliates in eight Latin American countries in the forms of dividends, royalties/fees, and interest payments, as well as a measure of intra-company indebtedness.

Notice that dividends clearly rank as the most important means of remitting funds, among these three methods. Measures of pricing strategies, swaps, and other transfer methods are not available at this level of aggregation. These results cannot be evaluated directly in the context of Table 10.3, because a major determinant of the choice of remittance vehicle is national law in the host countries. For example, the Andean Pact countries (Bolivia, Colombia, Ecuador, Peru, and Venezuela) do not allow royalties to be paid to foreign parent firms by their local affiliates.[6] However, a contingency table comparing the use of royalties relative to total funds remittances for Andean Pact countries versus other Latin American countries showed no significant difference in such transfers. Grosse (1983) did find that in the three years immediately following imposition of the Andean regulation (in 1971), there was a significant decline in the use of royalty payments by affiliates in those countries compared to others in Latin America. Apparently, that decline had been reversed by 1983.

Corporate strategies

An attempt to measure the difference between use of loans and royalties was not feasible, since no data were available on gross loans from affiliates to parent companies. Also, the other, untaxed methods may be used widely, but no aggregated data is available to confirm this point.

Table 10.4 Remittances from US company affiliates in Latin America, 1983($m)

Country	Dividends	Interest payments	Royalties and fees	Sum of Div + int + royalty	Net debt of parent to affiliate[a]
Argentina	234 (81%)	10 (3%)	45 (16%)	289 (100%)	−10
Brazil	402 (80%)	65 (13%)	35 (7%)	502 (100%)	107
Chile	39 (61%)	6 (9%)	19 (30%)	64 (100%)	−13[b]
Mexico	135 (40%)	72 (23%)	114 (37%)	311 (100%)	191
Colombia	121 82%)	7 (5%)	20 (14%)	148 (100%)	53
Ecuador	22 (73%)	1 (3%)	7 (23%)	39 (100%)	0
Peru	(D) −	(D) −	17−	−	107
Venezuela	129 (87%)	12 (8%)	7 (5%)	148 (100%)	−17

Source: *Survey of Current Business*, August 1984, and unpublished data from the US Department of Commerce
Notes:
[a] This column shows net borrowing of US parent firms from their Latin American affiliates, including trade credit.
[b] Negative values mean that the affiliate has net indebtedness to the parent company. Positive values mean that the affiliate is lending to the parent.
(D) means that the data were suppressed to protect company confidentiality.

Interview data

To view the situation from another perspective, consider the responses of MNE financial managers who are dealing with affiliates in the region. In summer of 1984 a questionnaire was mailed to thirty-two Chief Financial Officers for Latin America in US MNEs. These firms were chosen because they have Latin American regional offices in or near Miami, Florida — though many of the CFOs were located at the home office. No examination is made of the bias in this sample. The results presented here are intended simply to show tendencies in behavior among large US-based multinationals that have affiliates in Latin America. (The firms are all manufacturers, with the exception of two oil companies; and all had annual sales in excess of $1 billion in 1983. Their affiliates are virtually all subsidiaries, and there is a wide range of ownership patterns, from minority joint ventures to wholly-owned subsidiaries.)

Across the sixteen respondents to the survey, strategies for funds transfer were extremely varied. Looking first at methods for initially financing foreign direct investment projects in Latin America, 27 percent of the respondents replied that they use direct funds transfer from the parent for all or part of the capital base, 47 percent responded that they utilize local borrowing for all or part of their initial investment, while only 20 percent use foreign borrowing from either another affiliate or a bank.

Well over half of these firms (67 percent) do make intra-corporate loans to affiliates in the region, with only 7 percent of them sometimes using loans from the Latin American affiliate to the parent. Most of the respondents (67 percent) structure intra-firm loans to allow leads or lags in payment, that is to adjust the timing of repayment, for strategic purposes. Four-fifths of the firms prefer loans to equity transfers for funding Latin American affiliates, presumably because the loans guarantee repayment that host governments generally honor, in comparison with profit remittances, which often are limited. The large majority of firms (67 percent) charge interest rates above a base of LIBOR or US prime, much as do bank lenders.

More than half (53 percent) of the respondents transfer significant amounts of products/parts to their affiliates in Latin America. Of these firms, 57 percent stated that they use a strategy of arm's-length pricing for these shipments, while the others use negotiated prices, cost-plus, or some other method. Responses to this type of question are quite sensitive, because both host and home governments have been criticizing multinational enterprises for using 'unfair' high or low transfer prices. Thus managers can be expected to claim the use of arm's-length pricing, regardless of a firm's actual strategy. No conclusions about pricing strategy could be drawn from the set of questions aimed at this issue in the survey.

Some 69 percent of the respondents charged royalties on intra-conpany transfers of technology. Royalty rates generally were based on sales of the affiliate, with an average of about 4–5 percent of sales. A somewhat smaller percentage of firms (50 percent) responded that they charge management fees to their affiliates. No average rate was measured. Only 21 percent of the firms stated that they were able to time (i.e lead or lag) the royalty/fee payments as a strategic tool.[7]

Other funds-transfer methods mentioned by the respondents included: (1) parallel loans; (2) letters of credit (which allow access to foreign exchange); (3) discounting host government bonds offshore (in the case of Argentina); and (4) engineering fees. Overall, the major methods cited in the survey repsonses were those discussed here — loans, dividends, fees, swaps, and transfer pricing.

Conclusions

Considering only tax constraints on financial transfers from MNE affiliates in Latin America to parent firms, the choices can be evaluated fairly clearly. On that basis, Table 10.3 presented optimal transfer methods for use in eight countries of Latin America. When additional government policy constraints are taken into account, the choices are much more limited, though the same reasoning can be applied. In the early 1980s the Latin American debt crisis has resulted in severe limitations on funds transfers from the countries of the region, complicating matters further and encouraging the use of other methods to transfer *value* to or from desired affiliates.

Each country in Latin America has rules on financial transfer that differ from those of other countries. No single corporate strategy will serve in all countries, though the general idea of minimizing transfer costs can be followed in each case. Because well-defined limits exist on the traditional methods of funds transfer in most countries of the region, financial managers need to consider innovative strategies. Inter-firm swaps of various types avoid the immediate problem of foreign-exchange unavailability, and they may offer an important tool to the MNE financial manager in Latin America.

Finally, one should recognize that this entire analysis is done in a partial equilibrium framework. That is, it assumes that prior decisions have dictated the direction and volume of funds transfers needed within the MNE to optimize funds' uses and to protect against exchange risk. Those decisions themselves can conceivably be made simultaneously with the transfer decisions to maximize the firm's value — and they need to consider future, dynamic strategies of the firm, not only the goal of returning funds to the parent in the current time period.

Notes

Acknowledgements: This article is reprinted with permission from *Management International Review* (1/1986). Thanks to Don Kruse, who wrote an earlier paper that led to this project, and to the University of Miami for research support that enabled the author to perform the analysis.

1. An analogous equity transaction would be an investment by the affiliate in shares of the parent firm. The US Internal Revenue Service would construe this as a dividend, and most likely the host government would do so as well.
2. Rutenberg (1970), p. 45, discusses some similar issues regarding loans within his model of financial flows.
3. However, if a dividend is stipulated in the contract (e.g. dividends on preferred stock), it can be treated as an interest payment, but without the benefit of a host-country tax deduction. Also, if the affiliate is established

as a *branch*, rather than a subsidiary, the earnings will be taxed in the US currently, whether or not they are remitted.

4. Dumping is a very real concern in the United States today, as Japanese, Korean, Brazilian, and other firms are being accused of exporting steel at unfair prices, and many other firms of similar practices in other industries. Literally hundreds of cases are now being investigated by the US International Trade Commission for possible sanctions on the foreign exporters.

5. It is conceivable that a low transfer price plus a larger royalty payment would result in a lower tax burden than a high transfer price plus a royalty for transferring the funds which do arrive in the subsidiary.

6. Many exceptions have been allowed by the individual countries, hence the royalty payments to parent companies have continued to occur, though at decreased levels.

7. Interestingly, 50 percent of these firms license outside companies in Latin America to use their technology.

Bibliography

Batra, R.N. and J. Hadar (1979) 'Theory of the multinational firm: fixed versus floating exchange rates,' *Oxford Economic Papers*, July.

Bergsten, F., T. Moran and T. Horst, (1978) *American Multinationals and American Interests*, Washington, DC: Brookings.

Booth, E.J.R., O.W. Jensen (1977) 'Transfer prices in global corporations under internal and external constraints, *Canadian Journal of Economics*.

Chudnovsky, D. (1982) 'The changing remittance behavior of United States manufacturing firms in Latin America, *World Development*.

Cline, W. (1983) *International Debt and the World Economy*, Washington, DC: Institute for International Economics.,

Copithorne, L.W. (1971) 'International corporate transfer prices and government policy', *Canadian Journal of Economics*, August.

Grosse, R. (1983) 'The Andean Foreign Investment Code's impact on multinational enterprises,' *Journal of International Business Studies*, winter.

(1985) 'An imperfect competition theory of the multinational enterprise,' *Journal of International Business Studies*, spring.

Horst, T. (1971) The theory of the multinational firm: optimal behavior under different tariff and tax rates', *Journal of Political Economy*.

Kopits, G. (1971) 'Dividend remittance behavior within the international firm: a cross-country analysis', *Review of Economics and Statistics*, August.

Kopits, G. (1976) 'Intra-firm royalties crossing frontiers and transfer pricing behavior,' *Economic Journal*, December.

Lall, S. (1973) 'Transfer pricing by multinational manufacturing firms,' *Oxford Bulletin of Economics and Statistics,* August.

Merville, L. and W. Petty (1978) 'Transfer pricing for the multinational firm,' *The Accounting Review*, October.

Obersteiner, E. (1973) 'Should the foreign affiliate remit dividends or reinvest?' *Financial Management*, spring.

Rutenberg, D. (1970) 'Maneuvering liquid assets in a multinational company: formulation and deterministic solution procedures, *Management Science*, June.

Part four

Some Illustrations of MNEs in Latin America

Chapter eleven

The chemicals industry in Latin America

Introduction

General characteristics of the industry

The chemicals industry in Latin America is comprised of multinational firms as dissimilar as EXXON and Borden, and local firms ranging from tiny family-owned importers of fertilizers or paints to huge state-owned giants that produce petrochemicals and preclude foreign firms from some market segments. All of the world's twenty largest chemical producers have subsidiaries in Latin America, with concentrations in the largest markets. Many other chemical firms from industrial countries are active in the region as well.

The industry itself may best be viewed as a set of company groups, with relatively infrequent competition between competitors in more than one or two of the groups. There are producers of *'commodity'* chemicals such as sulphuric acid, ethylene, propylene, benzene, and urea, which are building blocks for other chemicals further downstream in the production process. In another group there are producers of *industrial chemicals* that use the basic chemicals as building blocks, but which themselves often are produced using proprietary technology. A third group are producers of *fertilizers and pesticides* for the agricultural sector. Another group produce various kinds of *plastics*. Yet another group, ignored in this analysis, are firms specialized in *pharmaceuticals*. This last segment of the chemicals market is the subject of such intensive scrutiny and continued controversy, that it is left for other sources to examine in detail.[1] Figure 11.1 sketches some of the basic chemical relationships between basic building blocks and downstream chemicals.

The unifying characteristic of all these businesses is a dependence on chemistry as the key technology underpinning the products. In terms of the industry in Latin America, most of the chemicals are petroleum- or natural-gas-based, that is petrochemicals. None the less, other non-petrochemicals are important in the business — for example ethanol produced from sugar cane and soybean oil.

Figure 11.1 A sketch of some key petrochemical production sequences

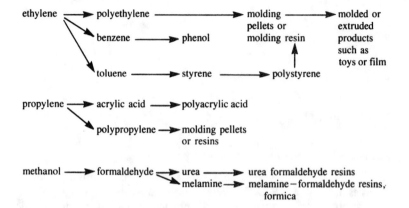

The leading role of MNEs in Latin America

Many, though by no means all, of the main competitors through the range of chemical businesses and countries in Latin America are the MNEs from the United States, the European Community, and Japan. Table 11.1 lists the twenty largest chemical companies in the world and notes some of their characteristics as related to Latin America. These companies generally overshadow local competitors in the segments that they serve in any Latin American market. The major exceptions are the petrochemical divisions of the largest state-owned oil companies such as Pemex and Petroquisa (chemical subsidiary of Petrobras), which also constitute major competitors in the region. Because so few of the government oil companies operate substantial chemical divisions, the competition in most chemical businesses is mostly among private-sector firms, particularly those listed above, but also including some large locally-owned firms in the larger countries.

It should be recognized that, while the large MNEs provide much of the competition for chemical products, an important part of the actual production of these products occurs outside of Latin America. That is, much of the sales that do occur in Latin American markets come from imported chemicals. Due to the large-scale economies that exist for many products, production tends to be centralized, and usually carried out in the firm's home market (most often the United States or Germany). Local sales offices and distributors are the most common form of MNE

Table 11.1 The world's largest chemical companies (ranked by 1985 sales)

Company (country)	1986 sales (in US$m)	Number of subsidiaries in Latin America	Main products made in Latin America
DuPont (USA)	27,148[b]	7	Agricultural chemicals; explosives; synthetic fibers; paints
BASF (Germany)	18,637	8	
Bayer (Germany)	18,765	8	
Hoechst (Germany)	17,506	8	
ICI (UK)	14,869	7	Pharmaceuticals; plastics; agricultural chemicals; dyestuffs
Dow (USA)	11,113	11	Plastics; agricultural chemicals; pharmaceuticals; specialty chemicals
Ciba-Geigy (Switz)	8,869	11	
Montedison (Italy)	8,609	8	
Rhone-Poulenc (France)	7,608	3	
Monsanto (USA)	6,879	7	
Akzo (Neth.)	6,373	5	
Union Carbide (USA)	6,343	8	
Fuji Photo (Japan)	4,064	n.a.	
Mitsubishi Chemical (Japan)	3,970	n.a.	
American Cyanamid (USA)	3,816	5	
W.R. Grace (USA)	3,726	7	Battery separators; industrial chemicals; sealants; silica; packaging
Hercules (USA)	2,615	4	Resins; plastics
Rohm & Haas (USA)	2,067	10	Agricultural chemicals; industrial chemicals plastics
Air Products & Chemicals (USA)	2,040	2	

Sources: Business Week, 20 July 1987; and 17 April 1987; company interviews; and *Moody's OTC Industrial Manual*, New York: Moody's Investors Service, 1987.

Notes:
[a] Number of Latin American countries in which the firm has a subsidiary.
[b] This includes sales of the Marathon Oil Company, acquired in 1983.

participation in the Latin American countries, though the largest markets have attracted major direct investments in production as well.

High-tech and low-tech segments

Not only is the industry multi-faceted in terms of product groups, but it also is quite varied in terms of technology intensity. Some of the base chemicals are considered very low-tech, requiring little R & D and competing primarily on price and availability grounds. At the other extreme, pharmaceuticals and some specialty chemicals require huge amounts of R & D (mostly done outside of Latin America) and are classified as very high-tech.

Even the high-tech product groups consist of production processes that range from standardized to highly innovative. In pharmaceuticals manufacture, for example, the basic research and production of chemical entities is high tech and capital-intensive. At the final stage of production, that is formulation of pills, liquids, tablets, and so on, the process is quite standardized, and requires relatively little capital. So, not only may *products* be more or less technology-intensive, their production *processes* also vary in technology intensity. As a consequence, the ability of Latin American governments to attract chemical production varies at the different stages of production.

Petrochemicals versus others

Petroleum-based chemicals constitute the largest part of the overall industry around the world in the 1980s, and the situation in Latin America is no exception. Multinational company estimates of the proportion of petrochemicals in total sales is about 80 percent, with much of the remainder still organic chemicals such as sugar-based alcohol derivatives and plant oils. For this reason, most of our discussion focuses on petrochemicals. As a point of reference, the major petrochemical products that constitute much of the total business are ethylene, propylene, ethylene oxide, benzene, toluene, styrene, phenol, butadiene, vinyl acetate, acetic acid, methanol, and ethanol (from sugar cane). These building blocks probably account for at least 75 percent of the downstream chemical products produced in Latin American countries.

The competitive environment

Aggregate direct investment in Latin American chemicals

A look at the size of the foreign chemical sector in particular countries may help illuminate the broad picture of competition in the region.[2] Unfortunately, very little data is available at this level of disaggregation except for US-based firms. Since the majority of direct investment in chemicals is owned by US firms (though at present the German and other European firms have about the same amount of total investment in Latin America), data on their FDI may serve as an acceptable indicator of the total. The US Department of Commerce publishes annual measures of the value of US direct investment in selected Latin American countries for the 'chemicals and related products' industry. These data for the past twenty years are presented in Table 11.2. The growth of FDI in the chemical sector is most impressive in Brazil and Venezuela, though in the other two large countries, Argentina and Mexico, growth also was greater than 10 percent per year in nominal dollar terms during the period before the debt crisis. Venezuelan growth certainly was fueled by the rapid rise in domestic demand for all products due to the oil boom of the 1970s. Although the foreign oil companies were nationalized in 1975, petrochemical firms were allowed to remain in business and serve the local market. Brazilian growth was fairly consistent throughout the period, dependent primarily on the growth of the huge national market, and surprisingly robust even in the face of the 1980s' debt crisis.

Table 11.2 US direct investment position in 'chemical products' at year-end (current $USm)

Country	1966	1970	1975	1980	1985	1986
Argentina	115	157	180	416	262	335
Brazil	99	210	528	1,036	1,324	1,402
Chile	13	13	(D)	23	23	26
Colombia	74	79	133	184	197	172
Jamaica	6	(D)	14	n.a.	89	74
Mexico	261	375	720	1,061	758	709
Panama	(D)	(D)	99	151	111	110
Peru	20	24	28	33	16	8
Venezuela	66	(D)	168	346	179	115
Other Central America	25	−9	60	129	75	70
TOTAL Latin America & Caribbean	781	1,128	2,176	3,594	3,141	3,101

Source: US Department of Commerce, *Survey of Current Business*, August issues, various years.

The major competitors

Most of the important multinational competitors in the Latin American chemicals industry were listed in Table 10.1. In addition, there are a handful of other firms that are diversified into chemicals but which are

213

generally classified into other industries or as conglomerates. For example, the major oil firms have quite large petrochemicals divisions, and both Shell and EXXON are very active in Latin American chemical markets. Also, firms such as FMC (a machinery, defense, and chemical company) and Borden (a dairy, food products, and chemical company) are key competitors in some chemical segments in Latin America. Finally, several of the multinational pharmaceutical companies such as Pfizer, Merck, and Hoffmann-La Roche operate numerous affiliates in the region and have extensive sales there. Despite the relevance of these firms to the total picture of the chemical industry in Latin America, our focus remains on the 'core' chemical companies, that is those that have most of their sales in industrial, agricultural, and commodity chemicals, and plastics.

Even with the limitations noted above, the overall industry must be divided into some categories of products, in order to give an idea of the kinds of competition that exist. That is, for example, Dow produces polyethylene but not Plexiglas, while Rohm & Haas makes Plexiglas and not polyethylene. The two firms compete in other products, but not these. Thus the competition must be defined to be among firms in each 'strategic group' of the total industry. The first cut at identifying such groups can be made according to product categories.

The main product groups, as noted previously, are:

1. Commodity chemicals, such as ethylene, sulphuric acid, urea, formaldeyhyde, benzene, styrene, chlorine, caustic soda.
2. Agricultural chemicals, such as fertilizers and pesticides.
3. Plastics, such as polyvinyl chloride, ABS, polystyrene, polyethylene, polypropylene.
4. Industrial and specialty chemicals, such as synthetic resins, leather-treating chemicals, dispersants.
5. Pharmaceuticals, such as ethical and generic drugs.

The main competitors in each category differ somewhat from country to country, but most of the foreign MNEs are active in at least the largest five or six countries of the region. Some examples of important competitors in the five product categories include:

1. Commodity chemicals: Dow (polyethylene, chlorine); ICI (Caustic soda, methyl methacrylate monomer);
2. Agricultural chemicals: ICI (pesticides); Rohm & Haas (fungicides); Bayer (herbicides); Dupont (crop-protection chemicals).
3. Plastics: Dupont (automotive parts); Rohm & Haas (Plexiglas).
4. Industrial chemicals: Dupont (Kevlar); Rohm & Haas (ion exchange resins); Dupont (titanium dioxide); Rhone Poulenc (dichloro analine dyes).
5. Pharmaceuticals: eg., Dow (Rifadin, Oteldane).

Competition by country

The extent of competition between the firms cannot really be seen through the product categories listed above, since the firms are not all active in all of the same markets, and local competitors are more important in some markets than in others. To improve the focus a bit, this section notes the country distribution of the major multinational competitors and their industry concentrations by country. The emphasis here is on local production by the firms, even though many product lines are supplied via imports.

Figure 11.2 is a map of Mexico, Central, and South America, with notations showing the locations of manufacturing sites and sales offices of the firms. The extent of affiliate networks roughly follows company size — the largest of the firms have the greatest number of affiliates in the region. Even in countries where a given company may not have an owned sales office or production facilities, there generally are distributors selling that company's (imported) chemicals. This means that, in fact, virtually all of the firms compete in all of the countries of Latin America.

A better idea of the companies' major commitments to Latin American markets can be seen from a breakdown of their main product lines sold in the countries where they operate affiliates. Figure 11.3 lists the countries in the region that are served by these firms and shows which are the main product lines sold by each competitor in each market. This view gives a better idea of the extent of head-to-head competition that takes place among the foreign MNEs, but it necessarily is still an incomplete picture. While agricultural chemicals are fairly commonly produced and sold by many of the companies in several Latin American markets, the degree of direct competition is not evident. For example, the herbicides produced by Bayer in Brazil do not compete with the fungicides produced by Dow there. Similarly, the broad heading of 'plastics' covers products sold by many of the firms in Mexico; but Dow's polyurethane does not compete directly with Rohm & Haas's Plexiglas. So, while the nature of head-to-head competition still is illuminated only imperfectly, we can see at least the kinds of products that are made locally by the major MNEs in each country.

The structure of the production process

Two of the fundamental characteristics of petrochemical businesses are that they tend to require up-to-date (sometimes proprietary) technology, and that they tend to require large-scale production at one or more stages of the production process. These factors lead companies to prefer to do much of their production in very large markets with good access to scientific resources and with low risk of political problems that could affect

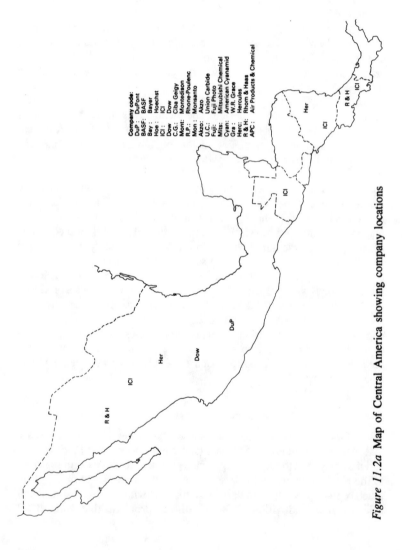

Company code:
DuP.: DuPont
BASF: BASF
Bay.: Bayer
Hoe.: Hoechst
ICI.: ICI
Dow: Dow
C.G.: Ciba Geigy
Mont: Montedison
R.P.: Rhone-Poulenc
Mon.: Monsanto
Akzo: Akzo
U.C.: Union Carbide
Fuji: Fuji Photo
Mits: Mitsubishi Chemical
Cyan: American Cyanamid
Gra.: W.R. Grace
Herc: Hercules
R & H: Rhom & Haas
APC.: Air Products & Chemical

Figure 11.2a Map of Central America showing company locations

Source: Company annual reports.

Figure 11.2b Map of South America showing company locations

Company	Venezuela	Uruguay	Peru	Panama	Nicaragua	Mexico	Guatemala	Ecuador	Costa Rica	Colombia	Chile	Brazil	Bolivia	Argentina
DuPont	i					p i / a				a		p i / a		i p
BASF	i					i a / f				s	s	i s / f		i a
Bayer						f						i f / a		i
Hoechst			i			p					i	a		i
ICI		f i				i f / a		f		c a	c	f		f p
Dow					a	a p / f c						a p / p c		f a / i p
Ciba Geigy	p f					i p					i p	i a / p f		i a / p f
Montedison						i p						f p		
Rhone-Poulenc	f		f			f								
Monsanto						i p						a p / s i		a p
Akzo												s		
Union Carbide								–		–		i c		
Fuji Photo (none)														
Mitsubishi Chemical												i p		
American Cyanamid														
W.R. Grace														
Hercules (none)														
Rohm & Haas						a p / i s			a s	a s		p a		i a
Air Products & Chemical												–		

Source: Company annual reports and interviews with Latin America division managers.

Figure 11.3 Chemical production in Latin America by major MNEs, 1987

their major facilities. None of the MNEs has a basic research facility in Latin America; such activity is typically limited to two or three locations in the United States and Europe. Most of the production facilities operated by these companies in Latin America are smaller than similar facilities located in the industrial countries, due primarily to the smaller market sizes (but also to some extent due to political risks).

In fact, the facilities operated by the MNEs in Latin America tend to be final finishing plants that 'formulate' the products sold to customers from ingredients made elsewhere, and sales offices that sell imported products to local customers. This point should not be overstated, however. In the large countries (Argentina, Brazil, Mexico, and Venezuela), most of the firms derive 70 percent or more of their sales from local production. It is the rest of the countries where most products are partially or totally produced overseas and shipped into the market for local sale. Another factor that importantly affects the MNEs' choices as to what to produce in Latin America is the existence of the large, government-owned oil companies. These state-owned firms preclude MNE entry into many market segments that otherwise would be attractive in the large Latin American countries.

Competitive advantages and competitive strategies of selected MNEs

This section sketches some of the characteristics of selected MNEs that compete in chemical businesses in Latin America. The firms were chosen simply on the basis of their availability to the author, and they are discussed in decreasing order of total global sales.

Dupont

Dupont has operated in Latin America since the 1920s, with manufacture of agricultural chemicals and plastics, as well as other products. Today the firm operates owned affiliates in fewer Latin American countries than most of its major rivals; Dupont's strategy at present is to work with independent distributors in most of the region, while maintaining manufacturing subsidiaries in the five largest markets. The company's single largest project is a titanium dioxide (paint pigment) plant in Mexico, which serves Dupont's needs for this product in Mexico and the rest of Latin America. Other important investments are in synthetic fibers (e.g. nylon, lycra) and industrial chemicals. Except in Venezuela, where the local content is somewhat lower, Dupont averages about 80 percent local value added in its sales in Latin America.

Dupont's competitive advantages in Latin America include (1) proprietary technology, especially in synthetic fibers and industrial chemicals; (2) product brands (e.g. Lycra, Dacron) that are associated

219

with high quality; (3) scale economies from its small number of large plants; and (4) its position as the low-cost producer of titanium dioxide in the world. As noted above, the firm's competitive strategy in the late 1980s is to operate affiliates only in the largest Latin American markets and to export to the others.

ICI

Imperial Chemical Industries (ICI) is as diversified and has as extensive operations in Latin America as its main competitors. ICI owns affiliates in nine Latin American countries, and it produces principally plastics and pharmaceuticals there. (See the list of product lines and countries of operation in Figure 11.3.) The firm entered Brazil and Argentina early, in joint ventures with Dupont. Its more extensive push into Latin America was relatively late, through its acquisition of US-based Atlas Chemical Company in 1971. Latin America constitutes about 4 percent of total corporate sales; about 70 percent of Latin American sales come from local production there. ICI's largest facility in the region is a new polyester film plant near Sao Paulo, Brazil.

ICI's main competitive advantages relative to the other MNEs appear to be (1) its access to large quantities of captive oil and gas, through its ownership of interests in the North Sea, and (2) its successful R & D record (ICI discovered polyethylene and developed polyester fibre). The firm has been less active in Latin America than its principal European rivals. ICI's broad competitive strategy in Latin America is to concentrate on its major business areas (e.g., agricultural chemicals, specialties, polyurethane, and polyester film) and develop them in size and competitiveness. More specifically, the firm is seeking to focus on product lines in which ICI is a leader and to 'grow' them further. Given the macroeconomic situation in the region, ICI has chosen to devote more of its efforts in the Big 3 markets (Argentina, Brazil, and Mexico).

Dow

Dow Chemical Company has the greatest amount of activity of the US firms in the region. With manufacturing in seven countries and sales offices in four more, Dow is as widely spread as any of its key competitors in Latin America. Different from most of its American rivals during the past decade, Dow has remained in the region with major emphases on both commodity chemicals and higher-tech products. The firm tends to choose major manufacturing investments in the larger countries in products that are not produced locally by any other competitor; then Dow fills out its product lines with imports and/or production from smaller facilities. For example, Dow's largest facilities in Latin America are its chlor-alkali and derivatives complex in Aratu, Brazil, its polyurethane, latex, and agricultural chemicals site in Argentina, and its plastics operations in Colombia and Chile.

Dow's key competitive advantages are viewed as (1) local production where Dow was able to be first in; (2) a better distribution network for agricultural chemicals (through independent distributors) than its rivals; (3) a balance between specialty and commodity chemical sales; (4) a strong marketing and sales organization throughout the continent; and (5) an outstanding know-how in finance (which is crucial in Latin America's volatile financial environment).

Dow's experience in Latin America is among the longest and most extensive, with plants in half a dozen countries by 1965. Dow's broad competitive strategy in Latin America is to invest wisely, capitalize on Dow's core technology, and operate with a long-term conviction that Latin America represents a natural high-potential marketplace and an attractive source of raw materials.

Rohm & Haas

Rohm & Haas Company is the 'youngest' of the companies discussed here in terms of its Latin American experience, having entered Argentina in 1949 and Mexico in 1951. Today Rohm & Haas operates in seven Latin American countries, with manufacturing in five of them. The firm's main businesses in the region include agricultural chemicals, plastics, and industrial resins. Its Plexiglas cast sheet plant (in the 'maquiladora' zone) in Matamoros, Mexico, supplies a portion of Rohm & Haas's sales for the western hemisphere. The firm's largest facility in the region is in Jacarei, Brazil, where Rohm & Haas carries out world-wide sourcing for agricultural chemicals and also produces specialties and ion exchange resins.

This firm has stayed away from production and sale of commodity chemicals, and has tried to position itself as a major competitor in specialty products. Rohm & Haas sees its competitive advantages in the areas of (1) technology leadership in its key products; (2) high quality as shown by its well respected trademarks; (3) technical service to customers in the region; and (4) its extensive organization throughout the region, especially in agricultural chemicals.

These sketches of company characteristics and strategies are intended to illustrate the nature of competition in chemicals businesses in the region. They also illuminate some of the complexities involved in discussing 'the chemical industry' in Latin America.

Relations with governments

Government relations play an important role in the operations of foreign firms in general and chemical firms in particular. First, the chemical industry is predominantly based on petroleum; and every major country in Latin America has its state-owned oil company that is a potential competitor, supplier, and customer for the foreign MNEs. Second, the pharmaceuticals business is so sensitive with respect to pricing, technology transfer, and other factors, that the chemical companies (that have pharmaceutical divisions) become subject to overall government scrutiny due to their participation in this segment of the industry. And finally, this industry is the source of large amounts of imported ingredients (and of potential product exports), which leads to serious government scrutiny for balance-of-payments reasons. Only two broad issues will be discussed here: the relation of state-owned oil companies to the chemical industry, and the bargaining relation between host countries and the foreign chemical firms.

Dealing with state-owned oil companies

Beyond any doubt the presence of government-owned oil companies in every medium-size and large Latin American country directly affects the petrochemical industry. For the most part these oil firms such as YPF, Petrobras, Pemex, and PDVSA, still specialize in the petroleum end of the market, leaving chemicals to the foreign MNEs and local private chemical companies. The state-owned oil firms have extended their activities downstream into refining of gasoline and a few other distillates, but they have largely avoided the technology-intensive and sometimes capital-intensive petrochemical segments.

None the less, the foreign chemical firms have encountered important limitations on their activities in oil-exporting countries. Both Mexico and Venezuela have imposed strict rules on investment in petrochemicals. Mexico, until 1987, required that all petrochemicals be sold through the state-owned Pemex. Venezuela, under Andean Pact rules, pushed foreign chemical firms to accept minority ownership in their Venezuelan affiliates from 1971–87. Since the beginning of the debt crisis in 1982, both of these countries and the others in Latin America have generally opened their economies to greater participation by these and other MNEs.

No simple conclusions can be drawn about the future of relations between state-owned oil companies and foreign MNE chemical firms. It appears very likely that, as Latin American scientists, engineers, and managers gain more and more skills, they will give the SOE oil companies greater ability to integrate downstream into petrochemicals. This situation will certainly lead to greater restriction of the foreign firms in the market segments selected by the local oil firms. The most likely candidates for expansion by Pemex, Petrobras, and so on, are downstream products based on their (already produced) basic chemical building blocks: ethylene, propylene, styrene, and butadiene. Some of these products are plastics such as polyethylene and polystyrene.[3]

Application of the bargaining theory

Based on the logic of the bargaining theory of government-business relations, one would expect greater regulation of the chemical companies' activities in those areas where the host governments have greater power in relation to the foreign companies. This kind of analysis requires the specification of those bargaining advantages and disadvantages possessed by each of the actors involved and the activities that may be affected.

The foreign multinational chemical companies possess three clear advantages in their bargaining with host governments in Latin America. First, they have proprietary technology that is used to create chemical entities and also in production processes. Second, they are able to achieve economies of scale in production that can be utilized through sales in several countries, not just the host country; so their costs would be lower in many cases than those of potential local producers. Third, they have information links to other countries, which may provide additional markets for sales, additional sources of supply, and managerial knowledge that may help to lower costs. On the basis of these advantages, one would expect that the companies would achieve better bargains with host governments when any one or more of these three factors is relatively important.

The host governments in Latin American countries have several key bargaining advantages as well. First and foremost, they have sovereignty over the national economy, so that they can determine the rules of the game for any business activity. Second, in general terms, the larger the market, the more attractive it is as a target for MNEs in the chemical industry. This implies that the larger countries should be able to place greater regulation on MNEs, which are more attracted to those markets. Third, the greater the part of raw materials in value added of a chemical business, and the greater the local availability of those needed materials, the greater will be the ability of the host country to regulate foreign MNEs. And finally, the more chemical companies that are able to supply a particular product, the greater will be the host country's ability to obtain a stronger bargain with the firm(s) that is allowed to serve the market.

While it is not easy to directly test hypotheses based on the advantages noted above, some preliminary analysis can be done using information collected in this study. The hypotheses for which tests could be done, or at least suggestive commentary made, are these:

H1. The greater the technology intensity of the product line, the more likely that line is to be imported and not produced locally.

H2. (a) The greater the production economies of scale for product, the more likely it is to be imported and not produced locally.

(b) The more capital-intensive a production process, the less likely it is to be carried out in a Latin American country.

H3. The greater the number of countries in which the firm does business, the less it is likely to be restricted by any given country, for example through local ownership demands, local content requirements, and/or price controls.

H4. The larger the host country market, the more restrictions will be placed on foreign MNEs.

H5. The greater the availability of locally produced oil, the more restrictions will be placed on foreign MNEs.

H6. The more MNEs offering to supply a particular chemical product, the greater regulation will be placed on that product.

The next six sections provide empirical evidence related to each of the hypotheses.

Hypothesis 1: technology intensity

The greater the technology intensity of the product line, the more likely that line is to be imported and not produced locally.

This hypothesis is appropriate for virtually any host country, except the country in which the firm initially produces the product (such as the United States or Germany). Greater technology intensity implies greater need for scientific and engineering monitoring and perhaps product adaptation, as market demands become better understood. These factors militate in favor of production in locations where the firm operates research facilities, namely the home country and occasionally one or two other industrial-country sites. By the same token, production would most likely be in the largest market(s), where customer demands can be seen immediately and responses formulated quickly.

In fact, what is typically seen when examining MNEs in the Latin American chemicals business is local production of a few commodity chemicals in large markets (e.g. Dupont's titanium dioxide in Mexico; Dow's polyurethane in Argentina) and importing of most other chemicals in finished or semi-finished form. What governments have been able to accomplish in their efforts to stimulate local manufacture by the chemical MNEs is to push them into setting up local *formulation* plants, which do final processing on chemical products to put them into the specific forms (such as specific shapes, states such as solid, liquid or gas, and sizes) that are sold to customers.

According to the managers interviewed in the sample firms, virtually no basic research is done in their Latin American affiliates. As a result of this and the relatively small market sizes, very little production of the basic chemical building blocks is done there either. Many of the firms do carry out laboratory and field development of agricultural products, such as applications of pesticides and fertilizers to particular climate conditions in different areas; so it cannot be argued that R & D is absent from the activities of the chemicals MNEs in Latin America.

The conclusion on Hypothesis 1, therefore, is that the technology intensity of a chemical product does appear to be inversely correlated with manufacturing investment by foreign MNEs in Latin America. In addition, high-tech chemical products generally have relatively low-tech final assembly stages, which have been attracted to Latin American countries, regardless of the technology intensity of the product itself.

Hypothesis 2: (a) Economies of scale (b) Capital intensity

> (a) The greater the production economies of scale for a product, the most likely it is to be imported and not produced locally.

This hypothesis also was tested only through interview discussions about the kinds of products produced in Latin American countries. A broad conclusion was that, in South America, transportation and regulatory conditions are so costly that products almost always are produced locally only if they can be sold in the local market. Therefore products requiring large volumes of output would not be produced in the region, or only produced in the largest market, Brazil. In Central America, on the other hand, the small countries individually justify production of very few products at all; but together they could attract some local production. More precisely, several companies have tried to use the Central American Common Market as a region for production in one country and sales in all five. With hostilities between El Salvador and Honduras and the Sandinista revolt in Nicaragua during the late 1970s, the common market effectively ceased to function, and the countries are viewed again as substantially five separate markets.

Given these logistical problems, it only makes sense for companies to restrict their production to products that do not require very large economies of scale, which in turn would require markets larger than most Latin American countries to absorb minimum efficient quantities of output.

> (b) The more capital-intensive a production process, the less likely it is to be carried out in a Latin American country.

This hypothesis is supported by the same evidence presented above, showing that the MNEs tend to site low-tech production in Latin America and to keep high-tech (and typically capital-intensive) production in the large, industrial markets. The final stage of the production process, formulation of the final chemical products that are sold to customers, tends to require less capital investment than basic production of chemical entities or basic research in chemistry. Thus there are many formulation facilities spread through the region, although still concentrated in the larger markets.

Another reason cited by managers in the interview sample for the MNEs' preference for less-capital-intensive investments is the political risk situation. In virtually every Latin American country, and especially the largest ones, the risk of currency inconvertibility is quite high. Due to the debt crisis, the likelihood of restrictions on financial transfers and imports of products and inputs is also very great. In all, the level of risk is relatively high in Latin America relative to industrial countries, therefore the companies prefer to keep their large, capital-intensive facilities in the industrial countries. Even in the instances where the firms operate large facilities in Latin America, these facilities tend to be less capital-intensive than similar plants in the United States or Europe. (This assertion is corroborated by evidence presented in Chapter 5 above, which explores the economic impact of foreign manufacturing investment on Venezuela.)

Hypothesis 3: Country diversification

> The greater the number of countries in which the firm does business, the less likely it is to be restricted by any given country, for example through local ownership demands, local content requirements, and/or price controls.

Because the companies are multinational and the governments are uni-national, it is expected that the companies have a bargaining advantage by not being hostage to any one government. As a hypothesis, therefore, we can see if a greater degree of multinationality correlates with a lower degree of government restriction. The measure used is the number of joint ventures in Latin America relative to the firm's total ventures in the region. We would expect that the firm would operate more wholly-owned ventures (and fewer joint ventures), the greater its number of affiliates in the region.

Except for the US-based firms, no data were obtainable concerning ownership of affiliates. Looking just at the US firms, it was found that the firms with more affiliates had greater propensity to use 100 percent ownership than the others. The Pearson correlation coefficient between number of Latin American affiliates and average percent ownership is 0.596, for the five firms on which data were available.

Measures related to local content requirements and price controls were found to be consistent across companies. That is, there was no difference in treatment of particular companies due to their degree of multinationality.

Hypothesis 4: Large market

> The larger the host country market, the more restrictions will be placed on foreign MNEs.

Clearly, one would expect that the more desirable the host country market, the greater interest the foreign MNEs would have in it. Consequently, the more desirable that market, the better bargaining advantage held by the host government. Market size (as measured by GDP or some other similar indicator) should correlate positively with the degree of restrictiveness of policy toward all foreign firms, including those in chemicals. A comparison country market size with average percentage ownership of MNE affiliates resulted in a Pearson correlation coefficient of $r = -0.411$. This shows the expected direction of influence, with larger countries having lower foreign ownership of FDI projects. A third comparison of country size with managers' perceived restrictiveness of the countries showed that, since the large countries have large oil companies, which in turn operate petrochemical subsidiaries, then those economies are more restricted to the MNEs than the smaller countries.

Hypothesis 5: Petroleum reserves

> The greater the availability of locally-produced oil, the more restrictions will be placed on foreign chemical MNEs.

Focusing precisely on the petrochemical industry, this hypothesis explores the idea that the greater the government's commitment to oil, the greater is likely to be its commitment to developing downstream activities in petrochemicals, and thus restricting foreign MNEs in those businesses. This is clearly true in the Mexican case, in which petrochemical investment by foreign firms is prohibited except by exception to the rules. Similarly, in Venezuela, foreign petrochemical firms have been held strictly to the Andean Pact rules on divestment of controlling ownership to local investors and otherwise limited in their activities. Beyond these anecdotal pieces of evidence, however, it was not possible to test this hypothesis.

Hypothesis 6: Product competition

> The more MNEs offering to supply a particular chemical product, the greater regulation will be placed on that product.

227

This hypothesis was most difficult to test, since the products in which there exists direct competition between or among MNEs are not that many, and because managers disagree to some extent on which products are competitive with others. Generally, managers' opinions were that this characteristic of Latin American markets was negligible in determining the degree of regulation.

Conclusions

The global chemical industry can best be understood as a group of sub-industries, each of which involves competition from a relative handful of major MNEs and often many local firms in any given country. The industry can usefully be disaggregated into segments of commodity chemicals, industrial and specialty chemicals, agricultural chemicals, plastics, and pharmaceuticals.

In Latin America the industry is characterized by competition from the major multinational firms, plus local firms in the largest markets. Most of the countries in the region import a large percentage of the chemicals provided by the MNEs. Argentina, Brazil, Mexico, and Venezuela have sufficiently large markets to attract large amounts of manufacturing investment by the MNEs, and most of the MNEs' sales in these countries come from local production.

The future of chemicals competition in the region is subject to great change in light of the strategies followed by the local government oil companies, which may very well expand their activities into downstream petrochemicals in addition to producing petroleum and refining fuels such as gasoline. At present, under the condition of external debt crisis, the foreign MNEs are encountering increasingly favorable treatment by host governments. These governments are greatly constrained in their policy options, since they need to attract as much foreign capital as possible to deal with debt servicing. When the debt crisis has been resolved, regulatory conditions may very well begin to become more restrictive, as in the late 1960s and early 1970s.

Notes

1. See, for example, Constantine Vaitsos, *Intercountry Income Distribution and Transnational Enterprises*, Oxford: Oxford University Press, 1974.
2. Even more preferable would be the size of the whole chemical sector in each country, if the data were available.
3. One MNE manager expressed the view that the Mexican and Brazilian national oil firms are more likely to concentrate on the basic chemical building blocks rather than expanding further downstream.

Chapter twelve

State-owned multinational enterprises in Latin America

Introduction

This chapter presents an overview of multinational firms that originate in the Latin American countries themselves. More specifically, the firms considered are those Latin American companies that are state-owned and which have established overseas operations of some kind. Similar to the foreign MNEs, the state-owned multinational enterprises (SMEs) tend to be the largest of the state-owned companies in the region. Also similar to the foreign MNEs, the SMEs tend to be concentrated in the oil, metals and chemicals industries and in other raw materials industries.

The present section describes some characteristics of the SMEs and lists the largest ones from selected Latin American countries. The second section looks at the firms' competitive advantages that enable them to succeed in international business. The third and fourth sections explore some of the reasons for setting up overseas operations and some of the alternatives to FDI for doing business abroad. The fifth section sketches some of the problems that many SMEs face and offers some limited evidence about their performance. The concluding section suggests what changes may take place in these firms' strategies in the years ahead.

The government sector and SMEs

State-owned firms can be defined reasonably consistently as government-owned business organizations that earn their revenues from sale of goods and/or services, maintain separate books from the governments themselves, and have a separate legal entity.[1] This definition distinguishes them from government departments that are less autonomous and often do not operate in markets for goods or services. Such firms are prevalent throughout Latin America in primary products (e.g. oil and copper) and public utilities (e.g. telephone and railroad service). These firms are run to varying degrees as profit-making companies, subject to greater or lesser government direction to serve social goals (such as increasing employment and locating facilities in underdeveloped areas of a country). Table 12.1 presents estimates of the size of government-owned companies as a part of the national economy in selected less developed countries. These numbers understate the participation of state-owned firms in each economy, since they do not include firms owned by local, as opposed to national, governments, nor firms minority-owned but still controlled by the government. Also, the figures understate the level of government participation in the economy, because they do not include the activities of the government itself, nor the banking sector, nor the military. In overall terms the government sector in every Latin American country accounts for over one-fifth of GDP, as was shown in Chapter 4.

Table 12.1 Nonfinancial SOE shares in GDP at factor cost

Country	Date of measure	%
Argentina	1975	8.6
Bolivia	1974–7	12.1
Chile	1982	17.7
Mexico	1978	7.4
Nicaragua	1980	36.0
Paraguay	1978–80	3.1
Venezuela	1978–80	27.5

Source: Data from Shirley 1983, 95.

The major state-owned enterprises (SOEs) tend to be concentrated in a small number of industries in each country. Historically, public utilities such as mail service, transportation, and communications were begun or nationalized by the state in Latin America. More recently, natural-resource companies in the oil, copper, tin, bauxite, and iron ore industries have been nationalized partially or totally in most countries. In addition, some major manufacturing industries such as steel, petrochemicals, and auto production have become targets of the move to state-owned business. The result of these public-sector inroads into the economy are that SOEs account for anywhere from 3 to 36 percent of GDP in Latin American countries. To give a more specific focus on these firms, Table 12.2 lists some of the most important Latin American SOEs (*not* all of which are multinational) in selected countries. The table has been constructed to show firms in the main industries populated by SOEs. In fact, these include all of the largest Latin American SOEs and most of the important multinational ones as well. Notice that every country has a national oil company, whether that country is an oil exporter or importer. The importance of oil to economic development in the twentieth century is such that none of the countries is willing to leave control

Table 12.2 Major state-owned enterprises in Latin America by industry

Industry	State-owned enterprise
ARGENTINA	
1. Petroleum	Yacimientos Petroliferos Fiscales (YPF); Gas del Estado
2. Communications	Empresa Nacional de Telecomunicaciones
3. Transportation	Ferrocarriles Argentinos; Aerolineas Argentinas
4. Electric power	Agua y Energia Electrical; Servicios Electricos del Gran BA
5. Metals	SOMISA (steel)
6. Banking	Banco de la Nacion Argentina
7. Other	CAP (Corporacion Argentina de Productores de Carne)
BRAZIL	
1. Petroleum	Petrobas
2. Communications	Telebras
3. Transportation	Varig
4. Electric power	Electrobras; CESP
5. Metals	Siderbras (steel); Cia. Vale do Rio Doce;
6. Banking	Banco do Brasil; BANESPA;
7. Other	Petroquisa (petrochemicals); Embraer (airplanes)
CHILE	
1. Petroleum	Enap (Empresa Nacional de Petroleo)
2. Communications	CTC; Entel
3. Transportation	Lan Chile
4. Electric power	Emelat; Endesa
5. Metals	Codelco (copper); Enaex (steel)
6. Banking	Banco del Estado
7. Other	Laboratorio Chile (drugs);
COLOMBIA	
1. Petroleum	Ecopetrol
2. Communications	Telecom
3. Transportation	Avianca; Empresa Colombiana de Ferrocarriles Nacionales
4. Electric power	Interconexion Electrica, S.A.
5. Metals	Ecominas
6. Banking	Banco del Estado; Banco de Colombia
7. Other	Carbocol (coal); Compania Colombiana Automotriz
ECUADOR	
1. Petroleum	CEPE
2. Communications	IETEL
3. Transportation	Ecuatoriana
4. Electric power	INECEL
5. Metals	Inemin
6. Banking	BEDE (national development bank)
7. Other	Compania Financiera Nacional
MEXICO	
1. Petroleum	PEMEX
2. Communications	Telefonos de Mexico
3. Transportation	Mexicana de Aviacion; DINA (Diesel Nacional)
4. Electric power	Compania Federal de Luz
5. Metals	Sidermex
6. Banking	Nacional Financiera
7. Other	Somex (financial and industrial conglomerate)

Some illustrations of MNEs in Latin America

Industry	State-owned enterprise
PERU	
1. Petroleum	Petroperu
2. Communications	EntelPeru
3. Transportation	AeroPeru; ENAFER (railroads)
4. Electric power	Electroperu
5. Metals	Centromin; SiderPeru (steel); MienroPeru; HierroPeru
6. Banking	Banco Commercial; Banco de la Nacion
7. Other	PescaPeru (fishing); ENCI (food importing)
VENEZUELA	
1. Petroleum	Petroleos de Venezuela (PDVSA)
2. Communications	CANTV (cia. Anonima Telefonos de Venezuela)
3. Transportation	Viasa; CAVN (Cia. Anonima Venezolana de Navigacion)
4. Electric power	Cadafe; Edelca
5. Metals	SIDOR (steel); Ferrominera del Orinoco
6. Banking	Banco de los Trabajadores
7. Other	Corporacion Venezolana de Fomento (CVF)

Sources: Business International Corporation, *Investing, Licensing and Trading Conditions* New York: BIC, 1987, various country pages; personal interviews.

of the industry to the private sector. Although there are quite a few SOEs in the petroleum and metals industries, and also in banking, none the less the governments do permit participation of foreign MNEs in most of these industries in most countries. For example, the multinational copper companies are allowed to operate in Peru, despite the fact that Centromin and MineroPeru own most of the ore reserves. Similarly, Texaco and other multinational oil firms are allowed to operate in some parts of the oil industry in Ecuador, despite the dominance of this industry by state-owned CEPE. In all, the presence of SOEs has not precluded foreign MNEs from most industries, though it does limit them severely and exclude them from some specific activities.

The main SMEs

The largest of the SOEs tend also to have the most overseas activities. Every one of the state oil companies listed above has several foreign subsidiaries. The international network of Petrobras affiliates leads the group, but Pemex and PDVSA also operate oil exploration affiliates, petrochemical plants, sales offices, and other facilities abroad.

Petrobras has invested in wholly-owned subsidiaries in oil exploration, production, and sales overseas at a value of over US$1 billion by 1985. Major affiliates exist in Algeria, Angola, China, Guatemala, Iraq, Libya, Nicaragua, and Trinidad & Tobago.[2] In addition, its trading

232

company Interbras has offices throughout Latin America, in the United States, and in Europe.

PDVSA, the Venezuelan national oil company, purchased a 50 percent share in Cities Service Company (Citgo), a major US gasoline station chain, in 1986. In addition, PDVSA has bought additional refining capacity in the Netherlands, by purchasing an oil refinery from British Petroleum. Also, PDVSA has formed a joint marketing venture with Champlin Petroleum Company in the United States.

In fact, the largest SOEs tend to be also the largest corporations in Latin America. Table 12.3 lists the largest fifty companies in the region in a survey carried out during 1985 for the magazine, *Progreso*. Note that the five largest firms are the oil companies in the largest countries, and that most of the rest are either subsidiaries of the largest industrial-country MNEs or other SOEs from Latin America. That is, there are relatively few locally-owned private-sector firms that belong to this group of the largest corporations in Latin America.

This perspective may be somewhat misleading, since quite a few industrial groups exist throughout the region that link companies through common individual or family owners but not corporate owners. For example, the Organisacion Diego Cizneros is the largest industrial group in Venezuela, but its main businesses (Pepsi Cola y Hit de Venezuela; Cada supermarkets; Venevision; Maxy's department stores) do not rank individually among the top fifty in the region. Before returning to the SMEs, let us look briefly at the industrial groups that provide important competition to both the SMEs and the foreign MNEs.

Large family-owned Latin American MNEs

An additional category of important MNEs that deal actively in international business in the region are family-controlled industrial groups such as the Polar Group in Venezuela and the Grupo Alfa in Mexico. A handful of the largest family groups appear in Table 12.3 above. (The family groups are marked with asterisks in the table.) The next few paragraphs sketch some of the characteristics of the largest family groups in Mexico, Argentina, and Venezuela.

In Mexico, Grupo Alfa has long been the largest of the family conglomerates. This firm and the Visa group belong principally to the Garza Laguera family of Monterrey. As a result of the debt crisis, this firm has undergone extensive restructuring and downsizing in the mid-1980s. Ownership shares have been taken by a number of foreign bank creditors whose loans were defaulted in the early 1980s. The group's main business are the steel manufacturer, Hylsa, and firms in petrochemicals and automobiles. Another key Mexican group is the Zambrano family, which owns the Cemex cement firm, Latin America's

Table 12.3 The largest fifty companies in Latin America

Ranking		Company name	Location	$M Sales		Assets	Profits	Number of employees	Main industry
1985	1984			1985	1984				
1	1	Petróleos Mexicanos	Méx	20.380,5	19.099,9	n.d.	15.145,0	n.d.	Petroleum
2	3	Petrobrás	Bras.	15.325,0	9.741,7	5.628,4	1.006,1	54.426	Petroleum
3	2	Petroven	Ven.	14.808,0	13.000,0	40.142,5	n.d.	n.d.	Petroleum
4	4	Y.P.F.	Arg.	4.903,3	2.605,1	9.234,5	679.8 —	32.216	Petroleum
5	5	Petrobrás Distr.	Bras.	4.107,0	2.600,5	388,8	20,7	3.964	Distribution petroleum
6	9	Shell	Bras.	2.673,0	1.604,4	462,4	2,5 —	3.104	Distribution petroleum
7	16	Volkswagen	Bras.	2.033,0	993,5	367,8	35,1 —	45.583	Vehicles
8	12	Esso	Bras.	1.882,0	1.227,0	164,2	20,1	1.410	Distribution petroleum
9	14	Souza Cruz	Bras.	1.698,0	1.138,5	107,4	56,2	14.898	Tobacco/Beverage
10	18	Pão de Açúcar	Bras.	1.661,0	884,7	179,9	18,4	5.200	Food
11	24	Ford	Bras.	1.585,0	807,2	154,7	27,9	22.000	Vehicles
12	17	Copersucar	Bras.	1.474,0	860,3	6,9	n.d.	3.178	Food
13	27	General Motors	Bras.	1.438,0	792,7	n.d.	n.d.	n.d.	Vehicles
14	11	Codelco	Chile	1.432,6	1.336,1	3.757,9	150,4	25.100	Mining (copper)
15	29	Vale do Rio Doce	Bras.	1.399,0	768,5	2.659,4	342,5	22.472	Minerals
16	37	Embratel	Bras.	1.345,0	627,4	938,0	172,6	11.452	Telecommunications
17	8	Ecopetrol	Col.	1.327,4	1.656,1	2.036,7	147,6 —	9.904	Petroleum
18	22	Atlantic	Bras.	1.299,0	810,6	59,5	9,0	1.431	Distribution petroleum
19	21	Texaco	Bras.	1.284,0	815,0	109,9	18,2	1.343	Distribution petroleum
20	34	Usiminas	Bras.	1.235,0	695,6	663,3	41,9 —	14.798	Steel
21	39	CSN	Bras.	1.176,0	604,4	1.080,4	447,4 —	22.691	Steel
22	35	Electropaulo	Bras.	1.150,0	667,6	339,5	13,5	20.521	Electric power
23	15	ENAP	Chile	1.137,7	1.084,6	964,6	9	4.087	Petroleum
24	41	Varig	Bras.	1.034,0	585,6	286,0	72,5	19.383	Airline
25	38	Petr. Ipiranga	Bras.	1.033,0	626,4	100,3	9,5	1.406	Distribution petroleum

26	19	Petro Perú	Perú	984,0	844,2	271,6	59,1	11.094	Petroleum
27	47	Cosipa	Bras.	947,0	549,8	1.027,7	157,3–	14.946	Steel
28	13	Federacafé/Fondo	Col.	936,5	1.216,0	1.290,9	446,6	5.323	Coffee
29	23	Esso	Arg.	934,7	809,8	n.d.	31,7	n.d.	Petroleum
30	33	Gas del Estado	Arg.	920,4	703,3	3.446,1	10,3–	9.723	Natural gas
31	30	Shell	Arg.	916,1	761,2	205,0	19,9	1.922	Petroleum
32	52	Pirelli	Bras.	912,0	528,8	286,3	46,4	12.262	Chemicals/petro-chemicals
33	55	CESP	Bras.	912,0	513,2	2.464,3	39,3	15.816	Electric Power
34	67	Mercedes Benz	Bras.	869,0	431,8	449,8	77,0	15.765	Vehicles
35	132	Mendes Júnior	Bras.	868,0	232,1	583,6	77,7	31.454	Heavy contruction
36	51	Interbrás	Bras.	865,0	529,8	195,4	25,9	1.800	Wholesale trade
37	74	Fiat Automoveis	Bras.	853,0	416,4	206,9	43,3	16.000	Vehicles
38	–	Org. Diego Cisneros	Ven.	833,0	n.d.	n.d.	n.d.	n.d.	Food
39	56	Copene	Bras.	819,0	493,6	845,4	46,9	1.591	Chemicals/petro-chemicals
40	26	COPEC	Chile	786,2	798,4	759,7	12,9–	520	Combustibles
41	31	General Motors	Méx.	766,1	745,3	737,9	n.d.	10.347	Vehicles
42	32	Chrysler de México, S.A.	Méx.	746,4	712,6	577,1	n.d.	10.483	Vehicles
43	59	RFFSA	Bras.	741,0	471,5	6.702,3	13,3	67.522	Transp. Ferrov.
44	87	Andrade Guitérrez	Bras.	737,0	342,2	385,4	92,5	23.674	Heavy construction
45	75	Rhodia	Bras.	719,0	411,4	279,8	52,1	9.636	Chemicals/petro-chemicals
46	63	Nestlé	Bras.	716,0	449,8	202,1	22,8	8.832	Food
47	20	Teléfonos de México, S.A.	Méx	709,4	824,3	3.037,8	n.d.	37.487	Communications
48	25	Cervecera Polar	Ven.	694,0	800,0	n.d.	n.d.	n.d.	Beverages
49	81	Agr. Cotia	Bras.	676,4	385,0	92,5	5,0	9.627	Agriculture
50	66	Camargo Corrêa	Bras.	675,0	432,2	715,9	78,7	20.025	Heavy construction

Source: Progreso December 1986.

largest. This firm has several direct investments in US cement production. A third and final example of Mexican groups is the Senderos family, which controls the Desc company. This firm is involved in a number of high-tech businesses, through joint ventures with Monsanto, General Electric, Merck, and others. In addition, Desc carries out extensive R & D on its own in chemicals.[3]

The largest Argentine family group is Perez Companc. This conglomerate is involved in the oil business, providing 10 percent of Argentina's total oil production. In addition, the group owns the construction giant Sade, which has direct investments in several other Latin American countries and which manufactures IBM computers under license. The bank associated with Perez Companc is Banco Rio, the largest domestic private-sector bank in Argentina. Even beyond these activities, the group is heavily involved in food businesses, with a major cattle-raising subsidiary and a major confectionary subsidiary, Aguila Saint. Perez Companc works with numerous foreign MNEs, as a partner in joint ventures (e.g. in Petroquimica Argentina with Uniroyal and Anglo American Corp.), as a licensee (e.g. with IBM), and as a business partner in various transactions.[4]

In Venezuela, the Organisacion Diego Cizneros ranks second in size among companies behind only PDVSA, the state oil monopoly. Its main businesses were noted above; in addition, the Cizneros group controls Radiovision — a chain of radio stations, Alimentos Yukery — a food-processing firm, and others. Cizneros is active in direct investment overseas, with extensive holdings in the United States (such as the Coca-Cola bottling company purchased from Beatrice Companies in 1986). The second largest group is Polar, which is controlled by the Mendoza family. This group's main business is the Polar brewery, which ranks fifteenth in the world. In addition, Polar owns most of Superenvases Envalic, the largest aluminum can producer in Latin America. The Mendoza family is also a large shareholder in Banco Provincial, the country's largest bank.[5]

These brief commentaries demonstrate the importance of the family groups in Latin American business, and they underline the relatively small size of most groups, which (except these few) are usually dwarfed by the SOEs and foreign MNEs.

The raison-d'être of SOEs

To understand better the decision-making process in state-owned enterprises, and especially those that operate multinationally, it may be useful to consider the reasons for their formation. Although most of them in Latin America have been formed since the Second World War, the basic logic behind the use of SOEs is the same in the older firms in

Europe and the newer ones in less-developed countries.

Generally, the most important factor in justifying the creation of an SOE is *revenue*. The government needs revenues to carry out its activities, and SOEs offer an easily tapped source of such income. Coal mines, copper mines, and later oil wells all offer readily-controllable sources of income for the SOEs (as well as tax income for the government itself). Additionally, some newer kinds of business have become the province of SOEs. For example, many countries have established 'marketing boards' for selling agricultural products.[6] These boards buy from individual farms and sell in international markets, generating revenues for the SOE via its price mark-up.

A second reason for the creation of SOEs is to preserve *sovereignty*. Governments always seek to maintain control over the functioning of the economy. If large corporations gain control over the most important industries, host country control may be difficult. Hence, often the largest industry (such as the oil sector) is state owned, as are the public utilities and some areas related to national defense. Governments also frequently try to keep ownership of key industries in domestic hands; so foreign MNEs often are targets of nationalization due to the attempt to retain national control over their businesses.

A third rationale for maintaining SOEs is *ideology*. Governments may view social rather than private-sector ownership of some industries as more in line with the responsibilities of the government to protect its citizens. Of course, in a socialist system, state ownership of business is a goal in itself. Toward the goal of improving income distribution, governments have turned to SOEs as a means of forcing redistribution on the economy.

The competitive advantages of Latin American SMEs

Those state-owned enterprises in Latin America that have entered international business and become multinational enterprises face formidable obstacles once they leave their home countries. For example, they generally have less technological skills than the private MNEs from North America and Europe, and they do not have government protection as do local firms in other Latin American host countries. Once exposed to competition from the large industrial-country firms and local SOEs in host countries, Latin American SMEs need some competitive strengths to offset their relative weaknesses. This section discusses a number of the key competitive advantages that exist for many of the Latin American SMEs.

Access to natural resources

The most important advantage held by the largest Latin American SMEs is access to and control over natural resources such as oil, copper, and coal. Ownership of a scarce natural resource provides the firm with an unassailable position at the initial stage of the production process. And with many minerals and metals facing growing demand as the LDCs industrialize and the industrial countries grow, the competitive strength of the resource-based SMEs will continue to grow.

What these SMEs have not possessed in the past is ready access to industrial-country markets for sale of their raw materials or downstream products that they yield. Historically, firms such as PDVSA and Pemex have relied on Exxon, Shell, and Texaco to market their crude oil in the United States, Europe, and elsewhere. Since the early 1970s, these firms have been expanding into distribution of crude to industrial countries, direct investment in oil refining in industrial countries, and even petrochemical production there. Latin American SMEs in the copper, aluminum, tin, and coal industries analogously have expanded into distribution and direct investment into downstream processing of those raw materials.

Given these firms' lack of R & D capability relative to the multinational oil or other extractive MNEs, and their lack of knowledge of industrial-country markets, it is clear that the primary competitive advantage that enables them to enter foreign markets is possession of the raw materials.

Access to local markets

The second category of the largest SMEs are public utilities such as electric power companies, telephone companies, and public transportation companies. These firms 'compete' domestically under monopoly conditions, in which the government forbids other firms from entering the markets for power provision, telephone service, and so on. Most of the utilities do not compete actively in foreign markets, although those that do tend to be important competitors. In electric power provision, a utility usually is a net supplier of power to other power providers or a net purchaser of power from other utilities. Those that have a net surplus for domestic use find neighboring countries as purchasers of their available power. Those that encounter excess domestic demand beyond their generating capabilities typically contract out to neighboring countries to meet those needs.

Telecommunications service providers must maintain linkages with other countries in order to carry out international calls or other transmissions. These linkages may be contractual or in the form of direct investment in telephone switching equipment, and so on. Typically, since equipment is provided by industrial-country MNEs, the SMEs do not operate extensively in direct investment. That is, their advantages are mainly in having captive domestic markets, and overseas markets do not offer advantageous opportunities.

Public transportation companies such as railroads and buses generally are limited to domestic markets, though their non-transportation business (and equipment purchasing) may be more transferable. Of course, the main national airlines all have overseas offices and service facilities in at least several other Latin American countries and in the United States. For example, Aerolineas Argentinas maintains facilities throughout South America and in the United States, the United Kingdom, France, and Spain.

Even in the oil-importing countries such as Brazil, the state-owned oil company possesses great competitive strength. This is obviously not because of its petroleum reserves, but rather because of its control over access to the Brazilian market for oil and petroleum-based products. Thus, access to the domestic market is an important competitive advantage for any SOE (or SME) that sells domestically as contrasted with firms that exist solely to serve foreign markets.

Government subsidies

For all of the kinds of SMEs discussed in the previous two sections, one central competitive advantage in *international* markets is the availability of a government subsidy. Since these firms are not managed to maximize profits but rather to serve multiple government goals (that *do* include profits, but not exclusively), often subsidies are necessary to keep the enterprise operating in spite of losses. From Table 12.3 it can be seen readily that many of the SMEs show losses on their business activities. This outcome is not fatal as with a private firm, since the government can 'bail out' the SME financially. Such subsidization is regularly necessary, because the losses shown in Table 12.3 are common among SMEs.

Large size

In relation to most of their domestic competitors, many of the SMEs are quite large in terms of sales, employees, and other measures. This means that they are able to achieve economies of scale in many areas such as purchasing, distribution, production, and financing. Each of these scale economies in turn enables the SMEs to lower its costs relative to rival firms. This advantage generally is not important when dealing with MNEs, since many of the MNEs are even larger than the SOEs. In fact, in the list of the largest SMEs shown above, only the largest dozen or so of them would even be listed in the Fortune 500. And all of them except for the four main Latin American oil companies would be far down the list.

Since many of the overseas activities of the Latin American SMEs are in *other* Latin American countries, size generally is an important advantage. Relative to local firms in those markets, the SMEs rank very high and attain significant scale economies as discussed above. In fact, one of the research findings about MNEs from LDCs in general is that they tend to invest in other LDCs, where they have relative advantages much as the industrial country MNEs have advantages over most LDC firms.[7]

Other competitive advantages

When considering the ability of SMEs to compete in industrial country markets, one advantage that often exists is *lower production costs*. Given the low wage and other costs in most LDCs, local firms in such businesses as steel production, autos, and foods, can achieve lower production costs than their industrial country competitors. This type of advantage may be offset, however, by poor product quality, high delivery costs, and restrictive government policies on imports in the industrial nations.

In LDC markets, on the other hand, SMEs tend not to have cost advantages, except as attained through scale economies. In these markets the ability to produce *specialized products* that are particularly appropriate for the local conditions is another advantage. For example, very often small-scale production processes are used by LDC-based SMEs; and these processes are very competitive in small LDCs where large production runs cannot be used up (see Wells 1983). Similarly, highly labor-intensive processes tend to be competitive in LDCs, where wages are low and capital costs relatively high. In all, products and processes that are typically eschewed by industrial country MNEs provide market niches for the SMEs from Latin America.

Reasons for going international

Why do the SMEs enter international markets? The reasons are quite varied, from seeking new markets to seeking new sources of supply, very similar to the reasons that other MNEs have for going overseas. The overwhelming reason for most firms from most countries is to seek new markets. This section reviews a number of the main reasons cited by SME managers for the initial decision to look to foreign markets or supply sources.

To sell natural resources

When raw materials supplies exceed national demand, the opportunity for export sales arises. For many Latin American SMEs, this is exactly the reason for entering international business. Subsequent to the decision to seek foreign purchasers, the SMEs often choose to establish foreign sales offices to deal with the new customers. Pemex, PDVSA, and CEPE, the three main oil exporting firms, have set up networks of overseas sales offices to market their oil and petroleum products. Similarly, Codelco and Mineroperu (copper), Bolivia's tin company, and other raw materials firms have expanded abroad.

To sell services

In addition to the primary products ventures, many of the financial services SMEs have extended their activities to other countries. The largest national banks (see Table 12.2) almost all sell their services through branches, agencies, and/or representative offices in a handful of other Latin American countries and in the United States (typically in New York and/or Miami). Banco do Brasil (with seventy-eight overseas branches), Banamex, and Banco de la Nacion Argentina have extensive foreign branch networks throughout the western hemisphere and elsewhere in the world. These banks typically service the needs of home-country clients in the foreign countries and offer services to local clients in the foreign countries who seek to do business in the bank's home country.

In addition to financial services, SMEs in raw materials and manufacturing businesses have discovered the ability to sell technical services to foreign customers, especially in other LDCs. Both Petrobras and PDVSA have been successful in selling oil production technology throughout Latin America and in the Middle East. Similarly, the steel firms Sidor and Siderbras sell manufacturing technology to other steel companies in LDCs.

To sell manufactured goods

Finally, the SMEs in manufacturing industries generally go abroad to find new outlets for their products, mostly in other LDCs in Latin America. Embraer, for example, has competed very successfully in the 1980s in the world market for sales of small commercial and military airplanes. The national steel producers in Argentina, Brazil, Mexico, and Venezuela all do substantial business elsewhere in Latin America selling their products. Petrochemicals subsidiaries of the national oil companies, such as Petroquisa in Brazil, have begun in the 1980s to expand their exports into industrial country markets, after beginning in other Latin American markets in the 1970s.

Empress Brasileira de Aeronautica (Embraer) is a highly-visible example of SME successes in overseas sales of manufactured goods. This Brazilian aircraft manufacturer has sold billions of dollars of its jets and helicopters in the industrial countries and elsewhere during the 1980s. About half of Embraer's sales are to foreign customers (mainly in the United States), and the firm has two foreign subsidiaries in the United States and Europe. Embraer's particular market segment is for small planes (two to twenty passengers), both for general aviation and for military uses. One of the very attractive features of Embraer's planes (such as the Bandeirante) to foreign customers is the low-cost financing provided through the Banco do Brasil, another SME in Brazil. It has been estimated that Embraer may take as much as 40 percent of the market in the United States for general aviation helicopters by the year 2000.[8]

Alternatives to foreign direct investment

In evaluating the importance of the foreign activities of Latin American state-owned enterprises, it is useful to examine the relationships between foreign direct investment (which creates SMEs) and exports, licensing, and other non-equity types of contracts abroad. As with virtually any international firms, the SMEs' exports are much greater in value terms than their direct investments. And also similar to other firms active in international business, the SMEs do relatively little overseas contracting, although in some cases this is rapidly-growing segment of the business. The next three sections comment on the export and contracting activities of Latin American SOEs, as well as their overseas portfolio investments.

Exports

While no data are available to measure the aggregate of Latin American SOE exports, one can safely assume that they dominate the total exports of Latin American countries. This is simply because the main exports of most countries are primary products, and these products are generally produced by state-owned firms. Pemex in Mexico by itself accounts for about half of the country's exports; CEPE in Ecuador for about 60 percent; and PDVSA in Venezuela accounts for about 90 percent of that country's exports.[9] While the contributions of individual SOEs to national exports are seldom as striking as these figures, still the exports of Codelco in Chile, Ecopetrol in Colombia, and Vale do Rio Doce in Brazil dwarf the exports of any private-sector firms in those countries.

Licensing

Licensing of foreign firms, especially other LDC-based SOEs, has grown into more than an exceptional phenomenon; in the late 1980s, several of the Latin American oil companies sell technical and managerial skills to other LDC oil firms; and some of the electric power suppliers and telephone company operators do likewise. The value of licensing and other contractual agreements is very low in comparison with either exports of SMEs or their direct investments. However, this kind of business is continuing to grow, as LDC managers, scientists, and engineers become more skilled and better able to utilize the technology that formerly was the province of industrial country MNEs.

Portfolio investment

The immense stockpiles of 'petrodollars' earned by Mexico, Venezuela, and Ecuador during the 1970s after the oil shock of 1973–4 have led to huge flows of portfolio investment overseas (as well as huge investments in development of their domestic economies). To be sure, the control over these funds often has been lost by the oil SMEs and suborned by governed agencies such as the Central Bank or the Finance Ministry — but the foreign portfolio investment that results is the outcome of the SMEs' successful business activity. In fact, it has been calculated that the portfolio-type foreign investment, such as eurodollar deposits and holdings of US Treasury securities, far surpasses the direct investments of the Latin American oil companies.[10]

The performance record

Given the multiple goals of SMEs (such as income redistribution, added employment, and reduced inflation — in addition to generating profits), it is not surprising that these firms do not show an impressive performance record. In this section, data from two sources are presented to document the financial performance of some of the SMEs during the 1970s and 1980s.

World Bank data

In a study of the management of SOEs, Shirley (1983) found that, except in Guyana, SOEs in Latin American countries earned close to zero profits during the period 1978–80. These data are presented in Table 12.4 below.

Table 12.4 Profits and subsidies of Latin American SOEs (as a % of GDP at market prices)

Country	Time period	After-tax profits	Current subsidies
Argentina	1976–7	2.1	1.0
Bolivia	1974–7	2.4	0.1
Chile	1978–80	1.5	0.3
Colombia	1978–80	1.0	0.3
Guatemala	1978–80	0.3	0.0
Mexico	1978	1.5	0.9
Panama	1978–9	0.6	0.4
Paraguay	1978–80	1.4	0.1
Peru	1978–9	0.6	0.1
Uruguay	1978–80	2.3	0.3

Source: Data taken from Shirley 1983, p. 11.

Progreso data

More recent performance data are available in the *Progreso* study of the largest enterprises in Latin America. The SMEs that appear in Table 12.3 have an average profitability that is significantly lower than that of the foreign MNEs and local private-sector firms in the list ($t = -2.24$. This shows a negative correlation of return on assets for state-owned versus private firms that is significant at the 96 per cent confidence level).

Venezuelan SME data

Finally, data have been compiled recently on financial performance of twenty-one of Venezuela's largest SOEs. These data, in Table 12.5, show that these twenty-one firms produced consistent losses during the period 1978–82, with no trend toward improvement.

Table 12.5 Consolidated financial performance of twenty-one Venezuelan SOEs (in millions of bolivars)

Year	1978	1979	1980	1981	1982
Total revenue	12.47	14.66	18.82	22.32	21.45
Total expenditures	15.20	17.76	24.54	29.76	25.18
Cost of production	10.58	12.71	17.38	19.29	16.37
Administrative and selling costs	3.99	3.85	4.71	6.24	6.66
Financial costs	0.96	1.73	33.50	4.62	2.77
Other income and expense	−0.33	−0.53	−1.05	−0.40	−0.61
NET INCOME BEFORE TAX	−3.34	−3.11	−5.83	−7.51	−5.03

Source: Kelly 1985, p. 143.

In all, the performance results of Latin American state-owned multinational firms is fairly bleak. This is to be expected in the sense that these firms are not operated solely to maximize profits; but certainly they are not intended to be financial burdens on the governments.

Conclusions

The safest conclusions to be drawn from this chapter are that Latin American SMEs tend to be large companies in raw materials industries, which perform poorly in financial terms, and which appear likely to remain in business (often through government subsidization) for many years to come. The implications of this phenomenon are discussed next, followed by comments on the late 1980s trend toward 'privatization' of SOEs in many countries.

Implications of SMEs for international business

Due to the statist tradition in Latin America (discussed in Ch. 4 above), there is little doubt that SOEs, and especially SMEs, will continue to play an important part in the economies of all countries in the region. These firms will continue to restrict the opportunities for foreign MNEs in raw materials industries and in public utilities. This implication is not new and is hardly surprising — but given the great importance of exactly these industries to foreign MNEs historically, it bears repeating.

As the state-owned firms that have already embarked on international ventures extend their activities, and as additional SOEs enter international business, firms from industrial countries will find additional competitors in their home markets as well as in Latin America and in other third-country markets. This means that firms such as Embraer, Petrobras, PDVSA, and Pemex will likely become important competitors in the United States and elsewhere.

Privatization

One of the 'solutions' to the Latin American debt crisis that has been discussed more and more convincingly is the sale of some state-owned companies to foreign creditors such as commercial banks. This vehicle could be used to reduce the size of foreign debt and simultaneously to increase investment in Latin American economies. Such privatization could as well be extended to domestic investors (i.e. not just foreign creditors), as has been done in Chile in the late 1980s. The historical record from 1985-7, during which time the concept has taken hold, shows that many European SOEs have been sold to private-sector investors, but few Latin American ones. While the pressure to repay foreign debt remains, efforts to privatize more Latin American SOEs will continue.

Indeed, the Latin American record is not devoid of privatization efforts. Ecuador in 1986 began a major program to sell shares in thirty-five companies owned by a state-owned development bank. Venezuela later in the year opened the petrochemicals sector to foreign participation; this is a striking shift from the nationalist policy that had been in place since the late 1970s. Both of these policy changes predated the 1987 elimination of Decision 24 in the Andean Pact, of which both countries are members.

Notes

1. Mary Shirley, *Managing State-Owned Enterprises*, Washington, DC: World Bank, 1983, p. 2.
2. Annibal Villela, in Sanjaya Lall (ed.) *The New Multinationals*, New York: Wiley, 1983, p. 225.
3. See, for example, 'Mexico's industrial groups,' *Business Latin America*, 27 April and 4 May 1987.
4. See, for example, 'Perez company group rapidly expands leading role in Argentina's private sector,' *Business Latin America*, 27 July 1987.
5. See, for example, 'Venezuela's family groups,' *Business Latin America*, 22 and 29 June 1987.
6. Yair Aharoni, *The Evolution and Management of State-Owned Enterprises*, Cambridge, Mass.: Ballinger, 1986, p. 97.
7. See, for example, Louis T. Wells Jr, *Third World Multinationals*, Cambridge, Mass.: MIT Press, 1983, Chs. 1 and 2.
8. Ravi Ramamurti, 'Las empresas del estado y la exportacion de productos de alta tecnologia: un exitoso caso Brasileno,' in Janet Kelly (ed.) *Empresas del Estado*, Caracas: IESA, 1985.
9. Data taken from International Monetary Fund, *International Financial Statistics*, August 1987, country pages.
10. The logic behind this result is explained very well in Kobrin, Steven, and Donald Lessard, 'Large scale direct OPEC investments in the industrial countries and the theory of foreign direct investment — a contradiction', *Weltwirtschaftliches Archiv*, December 1978.

Bibliography

Aharoni, Yair (1986) *The Evolution and Management of State-Owned Enterprises*, Cambridge, Mass.: MIT Press.

Kelly, Janet (ed.) (1985) *Empresas del Estado*, Caracas: Ediciones IESA.

Lall, Sanjaya (1983) *The New Multinationals*, New York: Wiley.

Shirley, Mary (1983) *Managing State-Owned Enterprises*, Washington DC: World Bank.

Wells, Louis T., Jr, (1983) *Third World Multinationals*, Cambridge, Mass.: MIT Press.

Chapter thirteen

The outlook for foreign MNEs in Latin America

Introduction

The intent of this book has been to offer an overview of conditions facing multinational firms that operate in Latin America, and to explore some of the strategies employed by these firms in the region. In this sense the previous chapters have described and interpreted the environment that is encountered by MNEs and the activities of the firms. The interpretation follows one central theme: a bargaining theory of the MNE. Using that conceptual framework, both government–business relations and general company strategies have been analyzed. The validity of the bargaining theory remains subject to much additional testing, and the theory itself requires much additional refinement; but the sketch provided here has offered a useful basis for understanding the Latin American business environment and the fit of MNEs in it. This concluding chapter suggests some trends that will affect MNEs in Latin America in the future and notes some additional issues which it was not possible to cover in the book. Let us first consider three overriding issues that will shape the environment facing foreign MNEs in Latin America into the 1990s.

The debt crisis

After more than six years of severe cutbacks in foreign private lending to Latin American governments and private-sector borrowers, the future for the next five years appears fairly easy to prophesy. Major banks have lost so much money in unpaid loans — some of which have been sold in the secondary market at discounts over 50 percent of face value, and others written down to similar values on the banks' books — that there is no likelihood that they will want to extend much new credit to these borrowers. Since the main group of borrowers that have been unable to service their foreign debts to commercial banks are governments, this implies that those governments will need to look elsewhere for increased financial resources (e.g. to official lending agencies and intergovernmental organizations). Another major class of debtor, however, are private-sector firms in Latin America. There is no similar stigma attached to lending to this type of borrower. Even in the late 1980s the extension of credit to private banks and companies has resumed a path of growth that will help to support Latin American development. (Of course, the lending banks now are much more careful to examine the creditworthiness of individual clients and loan projects, and to assure collateral on their loans, than they were previously.)

The fact that foreign private-sector financial resources have not dried up completely for Latin American borrowers does not mean that financial conditions are good or even satisfactory for the region. The economies of most Latin American countries are growing at less than 4 percent per year in real terms — not enough to help them substantially narrow the gap between the industrial countries and themselves. The regional depression of the early 1980s was sufficient to drop per capita national incomes back to levels of the early 1970s, and only now are incomes rising above those levels. Thus the last stages of the debt crisis still are creating a drag on growth that Latin American nations can ill afford.

Compounding the problem of reduced bank lending, foreign multinational companies have drastically reduced their rate of investment in Latin America during the 1980s. Because of limitations on profit remittance and even on access to foreign exchange for reasons other than profit remittance, these MNEs have chosen to constrain their own commitments of new funds to their affiliates in the region. Most of the new investment in the region during the 1980s has been a result of reinvestment of profits — often because the profits could not be remitted without severe loss of value (i.e. through controlled foreign exchange markets or taxes, either of which lower the amount of foreign exchange received at the parent).

By 1988 there appears to be some increasing confidence in the viability of projects in Latin America, so direct investment is growing once again. At exchange rates which currently do not excessively overvalue most Latin currencies, foreign MNEs are finding the environment somewhat attractive for new or expanded commitments, especially in manufacturing and services. To keep this point in perspective, it should be noted that a far greater amount of domestic and foreign direct investment has taken place in the United States in the late 1980s, as a result of that country's dynamic market and a devaluing dollar relative to European and Japanese currencies. Latin America represents such a small part of most MNEs' overseas activities that even a wave of greatly increased confidence in the region is not likely to attract a massive inflow of new companies and/or new facilities.

Public sector versus private sector

Despite today's relatively hospitable environment toward foreign MNEs, once the debt crisis has been overcome, a fundamental issue will arise again: what are the proper roles of public and private sectors in the economy? In a region where the statist tradition is very strong, and government-owned companies are powerful and widespread, the decision as to what part of the economy is to be left to the private sector and what part controlled by the government sector will often lead to new intervention in private markets. Since over one-fourth of the economy is government controlled (i.e. belongs to the government itself or to state-owned enterprises) in each country, clearly the SOEs have achieved important bargaining power with the governments that own them. Since difficulties in economic development often lead to attempts to blame scapegoat firms or industries, private sector firms are potential targets for such criticism. While there appears to be no trend toward eliminating the private sector altogether, still the possibility exists for a particular firm(s) to be forced out of a Latin American country due to the government's decision to control an industry segment.

Multinationals, of course, face this kind of threat even more than local private firms, because the MNEs can be accused of being foreign as well as private. Not only do the MNEs try to maximize corporate welfare rather than the country's welfare, but these MNEs also have corporate owners who are primarily foreign. Even without going to the extreme of nationalization, host governments may confront MNEs with increased regulation as they find additional bargaining advantages, and also when national economic conditions fail to improve as rapidly as the government views as acceptable.

SOEs remain concentrated in raw materials industries and public utilities, though they have increasingly pursued downstream business in their traditional industries. For example, several of the national oil companies have expanded downstream into refining and even into producing some petrochemicals. For most foreign MNEs, these state-owned enterprises constitute suppliers of production inputs and fairly ponderous bureaucracies — but not competitors. For the MNEs in the most capital-intensive industries, such as petroleum extraction, copper mining, and telephone service provision, the SOEs have largely forced out foreign owners. These MNEs have tended to be moved into the position of equipment and service providers and producers of downstream products in their core industries, as well as transporters of raw materials from the SOEs to overseas markets.

The legitimacy of foreign firms

The previous section focused on the division of an economy between public and private sectors. In that context foreign MNEs are one of many kinds of participants that make up the private sector. And the Latin American governments have created environments that range from Castro's Marxist Cuba to Pinochet's *laissez-faire* Chile, with many countries shifting alternately towards each extreme at different times. A second, related issue that separates MNEs from the rest of the private sector is their foreignness. It is a choice of each government as to what level of participation in the economy should be permitted to foreign firms and people.

In the final analysis, every national government has the sovereign right to decide what business activities are to be allowed and who is to be permitted to carry them out. As the example of Castro's Cuba points out vividly, even the right to do business in a Latin American country can be rescinded at the will of the government. Thus far, the same degree of nationalization of (domestic and) foreign business has not occurred in Nicaragua, but some companies have been nationalized almost everywhere in the region during the past two decades. The oil companies in Venezuela and the copper companies in Peru and Chile provide two striking examples. Given such events in the recent past, it is crucial to raise the issue of legitimacy of foreign firms.

Through careful, though none the less personal, observation, it appears to this author that the future economic regimes in almost all Latin American countries will continue to permit a major role for the private sector and for foreign firms. With the exception of Cuba, this is the case today. The possibility of a communist revolution in another Central American country or Chile exists, but does not appear probable. In the largest four economies — Argentina, Brazil, Mexico, and Venezuela — capitalism (which allows both state-owned companies and foreign MNEs to exist and compete) seems highly likely to endure indefinitely. These views tend to be borne out by the writings of other academic analysts as well.[1] Still, these are all assertions, and one cannot know if foreign MNEs will become unacceptable to some government(s) in some countries in the years ahead.

The direction of government–business relations

How well has the bargaining theory worked?

As an explanation of the interactions between host countries and foreign multinational firms, the bargaining theory has demonstrated an ability to shed reasonable light on a variety of phenomena. The relative success of high-tech firms in obtaining favorable entry and operating conditions in Latin America is one example. The relative lack of success of extractive firms in the oil and copper industries to maintain their earlier ownership positions is another. Based on the dozen or so bargaining advantages discussed in the book, companies and governments have interacted throughout Latin America during the post-Second World War era much as hypothesized.

Chapter 4 showed various measures of the significance of bargaining advantages in determining the government–business relationship in different countries of the region during the post-Second World War period. Not only was proprietary technology correlated with company strength and extractive industry correlated with government strength in their dealings, but several other hypothesized relationships were confirmed as well. Access to foreign markets has enabled offshore assembly plants in several countries — particularly Mexico — to maintain 100 percent foreign ownership and control and also to receive favorable tax treatment. Cycles in economic growth in the region have coincided with cycles in regulation of foreign MNEs: the higher the rate of economic growth, the higher the restrictions, and vice versa. Even advertising firms were seen to have relatively good bargaining ability compared to their counterparts in Latin America.

The bargaining theory that underpins the analysis throughout this book really is nothing more than a logical compilation of observed conditions in Latin American business in the 1960s, 1970s and 1980s. To say that in company–government negotiations, the party with stronger bargaining advantages obtains a relatively better share of the benefits of a business activity is not at all surprising. In this relatively uninteresting sense, it can be concluded that the bargaining theory has 'worked' in explaining some of the key aspects of foreign MNEs' activities in Latin America. The real test is in seeing whether the bargaining theory helps anticipate future conditions that will confront the MNEs and governments and whether it will show how to deal with those conditions.

One simple extension of the theory hypothesizes that, as SOEs and governments gain increasingly sophisticated information about production processes, marketing and management skills, and other technology, they will reduce the MNEs' bargaining advantages in those areas. This trend need not lead to total independence from outside suppliers of technology, but it can be expected to lead to stronger positions for governments throughout the region.

A correlated hypothesis is that governments will increasingly seek to obtain the advantages held by MNEs for themselves. To accomplish this, the governments are likely to try to force MNEs to sell proprietary technology and to rent or otherwise contract for additional advantages such as distribution channels to other countries and large-scale

251

production facilities. That is, the governments are likely to continue trying to eliminate the ownership and control that MNEs have traditionally maintained over subsidiaries around the world, and push the firms increasingly into being contractors for their needed services. This kind of outcome is already being seen in the constant pressures on foreign firms to accept local partners in their Latin American ventures, and particularly in the petroleum industry, where foreign firms have been nationalized at the production and refining stages in many countries and are left with equipment sales, technical service contracts, and marketing agreements.

Uneasy partners: the bargaining continues

In the late 1980s, under the cloud of the debt crisis, many examples can be drawn to illustrate the direction of relations between host governments and foreign MNEs in Latin America. This section notes five government policy changes in the context of the bargaining theory.

The Andean Pact's Decision 24

As discussed in Chapter 6, the Andean Pact's Foreign Investment Code was in place from 1971 to 1987. With its multiplicity of restrictions on foreign MNEs activities, it ranked among the most hostile of regulatory environments outside of the communist countries. Despite the antagonistic nature of the rules, several of the Pact's members permitted companies to avoid compliance throughout Decision 24's life, until in 1987 the rules were replaced by Decision 220, which eliminates the required fade-out of foreign ownership and calls for equal treatment of foreign and domestic firms in Andean countries. Almost continuously since the outbreak of the debt crisis, the Pact's Junta considered the elimination of Decision 24, until finally the lack of foreign capital and continuing economic recession led to its replacement with a policy designed to attract more FDI into the region.

Before the final repeal of Decision 24 but after the two oil crises, it appeared that not only were exceptions being made to high-tech firms that offered to bring in wanted scientific and managerial skills, but also Andean governments were trying simply to attract more capital. The MNEs do not offer the only source of foreign capital, but they do bring in investment that is not tied to fixed interest and principal payments (as in the case of bank loans). During the period of capital shortage in the region since 1982, MNEs have faced and may very likely continue to face reduced limitations on their investments, based on their bargaining advantage in access to financial capital.

Brazil's rule on informatics

The Brazilian government's decision to restrict access to the local market for computers in 1976 was the beginning of a long period of shifting regulations, varying degrees of limitations on foreign participation in the 'informatics' sector, and increasing friction between the United States and Brazilian governments on this issue. The restrictions on foreign MNEs became even more severe in 1984, when the Brazilian government denied access to the microcomputer segment of the market for all foreign firms. To date (late 1988), the foreign MNEs have only been permitted access to the markets for minicomputers and mainframe computers as far as hardware is concerned — and limits have been increasing on imported software in addition.[2]

This situation provides a marked counterpoint to the bulk of regulatory change during the debt crisis. While most countries are opening the doors to foreign direct investment and other forms of foreign participation in their economies, Brazil has significantly decreased foreign access to the informatics sector. The policy goal, of course, is to stimulate Brazilian computer scientists and entrepreneurs to develop and market home-grown microcomputers and software. This goal has been increasingly fulfilled since the original market-limiting policy was enacted in 1976. One can easily and accurately argue that these rules have slowed Brazil's access to the latest computer technology; but the result has not been devastating. And now, contrary to the other Latin American countries during the debt crisis, Brazil has *not* softened its anti-foreign policy on this sector.

In terms of the government–business relationship, it appears that the Brazilian government has asserted its strength due to the possession of the largest market in Latin America, and Brazil has forced foreign MNEs into limited segments of the computer market. Despite the technological advantages of firms such as IBM, Apple, Olivetti, and others, the Brazilian government has pushed the bargain against them. This outcome may become even more significant as other Latin American economies such as Mexico and Venezuela grow and become more attractive targets for sales by foreign MNEs.

Central American nations leave doors open to FDI

One of the clearest examples of a bargaining situation that favors the MNEs is that in all of the Central American countries. From Guatemala to Panama, each of the six countries is very small economically, troubled by political strife or even outright war, and highly dependent on US foreign aid for economic assistance. The situation is not a result of the debt crisis but rather it is a long-standing feature of Central America. The result of the weak economic environment is a general openness toward foreign investment in this region. The laws governing foreign direct investment in these countries are the most lenient in Latin America (perhaps with the exception of Chile).[3]

Even Nicaragua under the Sandinista government has not precluded foreign direct investment or nationalized the foreign firms already operating there. While maintaining a politically anti-United States stance, Nicaragua's government has not eliminated its private sector nor the foreign participants in that private sector.

Mexico permits exceptions to its joint-venture law

The Mexican government has not escaped the problems of other populous oil-exporting countries; it too has encountered a deficit in the current account balance of payments and a need to finance it with new international resources. As in the Andean case, one of the available sources of foreign funding is FDI, which has historically favored Mexico because of its proximity to the United States and its large internal market. Given its relatively desirable characteristics compared to other Latin American countries, Mexico since 1973 has been successful in dictating entry terms to foreign MNEs. The law requires foreign investors to find local majority partners for affiliates in Mexico. Exceptions have been few, beyond those firms already established in the country before 1973. In 1986, a very visible exception was made, when IBM was permitted to invest in a new personal computer plant without accepting local participation in its ownership.[4]

In addition, the entire phenomenon of 'maquiladora' plants has escaped local ownership requirements, as well as receiving other incentives to operate. These plants are used for offshore assembly in the low-cost labor environment provided by Mexico. From these locations, the American and other foreign firms are able to ship finished goods directly back into the United States through Texas and California, and end up with very low delivered costs. Despite its broad policy of forcing local majority ownership and control, Mexico has made notable exceptions for offshore assembly plants that number in the hundreds (and which employ several hundred thousand Mexican workers).

Chile's debt–equity swaps

Chile since the fall of the leftist Allende government in 1973 has followed the most open economic policies of any Latin American country. To attract foreign capital and direct investment after the nadir of 1973, Chile withdrew from the restrictive Andean Pact and implemented a very pro-foreign business policy in the following few years. Even these policies were insufficient in the eyes of government policy-makers during the current debt crisis, so Chile began a program of permitting conversion of foreign, dollar-denominated loans into equity in Chilean firms. This program attempts to attract new capital investment in the country at the same time as foreign indebtedness is eliminated through the swap of dollar loans for peso investments. (This idea was described in more detail in Ch. 7 above.)

During the first two years of existence of this program, Chile has converted over US$ 1 billion of foreign debt into local direct investment, and plans call for indefinite continuation of the swaps. Some of the investments have been in denationalized Chilean state-owned firms, but the vast majority are in private-sector ventures.

More restrictions in the 1990s

The above examples all point out the results of company-government bargaining under a situation of major government weakness. Continuing to follow the bargaining theory, it can be expected that, once the debt crisis has been overcome, the governments will possess even stronger bargaining positions relative to the foreign MNEs. In this situation, more restrictions on the MNEs are virtually assured. Once the severe capital shortage of the debt crisis has been overcome, host governments can be expected to chip away at the other advantages (such as scale economies and managerial experience), and accordingly reduce the MNEs' bargaining strengths. Also, the host governments will possess larger, more developed markets in the future — a major attraction to the MNEs, and thus a key bargaining chip for the governments.

A balancing item is the possibility of new bargaining advantages that the MNEs may develop. Certainly, the creation of new, sophisticated technology can give the possessor firm(s) a strong bargaining position. And access to industrial nations' markets will probably remain primarily in the hands of foreign MNEs rather than Latin American firms, local or multinational. Thirdly, the marketing advantage gained through development of successful branded products will not be neutralized in the region, except in cases where local MNEs acquire existing branded goods from industrial countries' firms.

Foreign firms and social and cultural issues

An entire area of MNE impact that has great significance to host governments has been excluded from the discussion to this point. Social and cultural effects of MNE activities undoubtedly have important consequences for the countries. Unfortunately, the tools of the economist or the business strategy theorist offer very little help in tackling these issues. Without going beyond a brief commentary, two types of impact that relate to sociocultural analysis are considered here: income distribution and 'Americanization' of the society.

Income distribution

Probably the greatest unanswered challenge to the record of MNEs in

Latin America is the criticism that they fail to improve income distribution in the host countries. While the foreign firms may contribute to the growth of GDP and/or exports, they generally do not have noticeable impact on the poorest segments of the population. In political science terminology, the people who belong to the 'center' of the host societies, that is those with the highest incomes, educations, and social status, tend to benefit from the participation of MNEs in the economy. On the other hand, people in the 'periphery,' who have very low incomes and little or no contact with external markets, are simply ignored by the MNEs. Since these people typically have inadequate incomes to buy the MNEs' products, the firms do not try to sell to them. Since these people usually are inadequately educated to work in the MNEs' factories or offices, they again are left out of the firms' activities. In sum, the periphery is often largely left out of the range of benefits generated by the MNEs.

Beyond the efforts of some of the firms and their individual managers to serve the poorest segments of host countries through contributions of their own time and money to charitable organizations, the MNEs do not deal directly with the problem of highly unequal income distribution in Latin America. A defense that is often made for the MNEs is that they pay taxes to the host governments, which in turn are charged with redistributing income justly within the countries. Blame can thus be passed to the governments. This defense is weak in that it fails to recognize that the MNEs have legitimacy in a host country only in so far as they are perceived to operate acceptably and serve the needs of the country. Thus, other methods of obtaining technology, financial resources, and so on, may be preferred to MNEs and FDI, if the MNEs fail to serve goals such as income redistribution.

An alternative perspective on distribution focuses on functional income distribution. Looking at the division of national income between factors of production — land, and capital — the MNEs' impact can be judged again. From this view, as seen in Chapter 5, direct investment tends to reduce returns to domestic capital and raise returns to local labor, which is generally considered to be a beneficial redistribution of functional income. This result should not be ignored, but it does fail to deal with the dramatic problems of the very poor in Latin America.

Cultural values: Americanization

A criticism of United States-based MNEs around the world has long been that they transfer American customs and styles to the rest of the world, often replacing local cultural patterns. While this transformation can be considered 'progress' in making more goods and services available to more people, it also serves to homogenize cultures and to place the host countries in a dependent position with respect to development of new styles and social traditions.

The criticism is valid to the extent that people around the world drink Coca-Cola, listen to American recording artists, watch American movies, and wear Levis. The fact that such styles have gained acceptance in dozens of countries does not demand that other societies follow the trend, but it does create a powerful demonstration effect that tends to result in more followers. In economic terms, there is no negative connotation of this phenomenon, especially if the products (such as soft drinks and clothes) are produced locally. In social terms, a case can be made for defending cultural traditions, and thus objecting to the imitation of foreign societies. Since the MNEs cannot avoid their cultural bases in the home countries, this problem will remain as long as international business takes place.

Many countries have responded to the Americanization of their societies by supporting local cultural activities and traditions. Governments operate or support museums, performing arts, and the like in order to maintain the cultural links to the domestic society. This often leads to additional difficulties, since most Latin American societies are descended from both indigenous Indian cultures and immigrant European (predominantly Spanish and Portuguese) cultures. These societies themselves are not fully integrated, and government responsiveness to either culture can be viewed as discrimination against the other. In all, there is not much agreement on a preferable alternative to Americanization — except that most people do prefer to maintain the pluralism inherent in European, Indian, and American populations.

Competition in Latin America

MNEs versus SOEs versus local family organizations

The competition between foreign MNEs and local firms is certainly bound to intensify in the years ahead. Both state-owned enterprises and family groups such as the Organisacion Diego Cizneros in Venezuela, the Miro Quesada group in Peru, and Grupo Alfa in Mexico will gain new competitive advantages through experience dealing with their rivals from industrial countries. Even now, the family groups often serve directly as licensees, franchisees, and joint-venture partners for competitors from the United States, Europe, and Japan. They are obtaining skills and knowledge that will enable them potentially to compete with their partners in the future. Of course, the MNEs develop new competitive advantages also, so the net result is not forseeable in precise terms.

Greater presence of European and Japanese firms

Another feature of the competitive environment in the next decade will be much greater participation by European and Japanese firms. Already Japanese banks have become a major force in lending and offering other bank services in the region. The Japanese electronics and auto firms, as in the United States, have established subsidiaries in the large markets and serve smaller markets through increasingly extensive distributor networks. The European chemical and pharmaceuticals firms already lead their American competitors in many Latin American markets, and more European firms can be expected since the dollar has been devaluing relative to European currencies since 1985.

Industry sectors

As in the rest of the developing world, foreign firms have been able to increase their activities in manufacturing and many services, while losing the opportunity to own extractive ventures and public utilities. Most of the public utilities were acquired by Latin American governments before the Second World War. Raw materials ventures, on the other hand, have been shifting to host government control slowly but surely over time. National oil companies have existed throughout the region for decades; Mexico even nationalized the foreign firms as early as 1938. It has only been since 1973 that Venezuelan and Ecuadoran oil production became primarily nationally owned. Copper companies were nationalized in Peru and Chile in the populist heyday period of 1965–75, but today several foreign MNEs operate mines and exploration ventures in both countries. Tin mines are largely nationally owned in Bolivia. The massive El Cerrejon coal project is jointly owned by Colombia's government and Exxon Corporation, with whole ownership reverting to the government in 2010. Reasonable speculation would say that the bauxite, iron ore, and other mineral industries, as well as lumber, fishing, and much of agriculture will be reserved for domestic ownership in the years ahead. Foreign multinational firms can expect to operate in Latin America in these industries through licensing, management, purchasing, and service contracts, as well as by selling necessary equipment.

Manufacturing remains a business that is relatively open to participation by foreign firms. The largest US and European manufacturers are well represented in Latin America with plants and factories, though most of them operate primarily in the largest three or four countries. From autos to appliances to food processing, world leaders such as General Motors and Toyota, Sunbeam and Sony, Kraft and Nestlé, maintain large market shares and some manufacturing operations in Latin America. Aside from a few industries that relate closely to raw materials (and thus would be likely candidates for expansion by local state-owned enterprises) such as petrochemicals, steel, and aluminum, most manufacturing appears likely to remain open to foreign competitors in the future. Perhaps the most likely significant change to be expected in the competitive environment in the near future will be the shift in ownership of firms from predominantly US parents to a much broader spectrum of European, Japanese, and North American (including Canadian) parents, as noted above. Given their technological and marketing successes elsewhere, the Japanese firms can be expected to expand their market shares more than firms from other countries.

In services, the field is too diffuse to draw any sweeping generalizations. Due to pressures from industrial-country governments with outstanding claims on Latin American governments, the financial services sector is likely to become more open to foreign competitors. This tendency is being reinforced as some US commercial banks seek to obtain equity positions in local financial (and industrial) institutions in exchange for non-performing loans. These are the 'debt–equity swaps' that have become quite popular in the past two years.

The industrial services sector, which includes licensing and management contracting, is expanding much faster than direct investment in any sector. This type of service is generally viewed as acceptable to Latin American governments and pressure groups as a means of gaining access to needed technology without giving away ownership and control to foreigners. In each of the nationalizations of foreign oil companies, service contracts have replaced foreign ownership; and the contracts continue to be renewed today. It is not much of a speculation to argue that this form of international business is very likely to increase greatly in importance in Latin America in the next few years.

The tourism sector, including hotels, airline and ocean transportation, and restaurants, remains fairly open to foreign participation throughout the region. At the same time, there is not extensive foreign ownership of tourist service provision in the region. For example, most of the hotel industry consists of locally-owned accommodations plus franchises of multinational hoteliers such as Holiday Inn and Sheraton. Restaurants are almost entirely owned by non-multinationals, except those in foreign-owned hotels. And airlines are highly restricted in their access to Latin American markets. In most cases there are only one or two US carriers allowed to serve each Latin American country from the United States, and an equal number of local carriers fly to the United States. In all, there is very limited participation of foreign MNEs in the tourism industry in Latin America.

Conclusions

A truly fundamental conclusion arising from the observation of foreign multinational firms in Latin America is that only a handful of them comprise the bulk of income, employment, exports, and other kinds of economic impact. If the activities of Exxon, Shell, Texaco (in oil), General Motors, Ford, Volkswagen (in autos), Anaconda, Kennecott, and Reynolds (in metals) are aggregated, a large percentage of total foreign business will be included. For example, in the smaller economies, natural resources make up the bulk of national income (ignoring the 'informal' sector.) When Exxon closed its refinery in Aruba in 1984, about one-third of GNP was lost. Similarly, when Shell terminated its refinery in Curacao, that country lost over 40 percent of its employment and income. In larger countries such as Brazil and Venezuela, the total impact of foreign firms is not so overwhelming, but the weight of few firms listed above in total *foreign* business still is quite high (i.e. over 25 percent of that total). Thus, despite the non-trivial activities of many other foreign firms, especially in manufacturing industries, a major part of total MNE affairs can be attributed to these few Fortune 100 firms.

In broad terms, the bargaining model gives a good idea about what we can expect as economic conditions change over time in Latin America. When the region is in expansionary periods, regulation on MNEs expands. When the region is in recession, rules toward MNEs soften and their bargaining capabilities improve. This characteristic of the twentieth century appears likely to continue into the next one, as long as governments are uni-national and companies are multinational.

Notes

1. See, for example, Guillermo O'Donnell, Philippe Schmitter, and Laurence Whitehead (eds), *Transitions from Authoritarian Rule: Latin America*. Baltimore: Johns Hopkins Press, 1986; also see, Wiarda, Howard, and Harvey Kline (eds), *Latin American Politics and Development*. Boulder, Colorado: Westview Press, 1985.
2. See, for example, Smogard, Gregory, 'Intellectual Property Rights Protection: The Case of the Brazilian Informatics Sector', University of Miami International Business & Banking Institute *Discussion Paper #88-3* (April 1988).
3. For a current overview of rules on foreign direct investment in Latin American countries, see Business International Corporation, *Investing, Licensing, and Trading Conditions*. New York: Business International Corporation (current edition).
4. Frazier, Steve, 'Mexico Hopes Its Approval of IBM Plant Encourages more Foreign Investment, *Wall Street Journal* (10 July 1985).

Bibliography

Business International Corporation, *Investing, Licensing, and Trading Conditions* New York: Business International Corporation (current edition).

Moran, Theodore (ed.) (1986) *Investing in Development: New Roles for Private Capital?* Washington DC: Overseas Development Council.

Wiarda, Howard, and Harvey Kline (eds) (1985) *Latin American Politics and Development* Boulder, Colorado: Westview Press.

Index